Synthesis Lectures on Technology and Health

The series publishes state-of-the-art short books on transformative technologies for health, wellness, and independent living. Our scope of publishing in the expanding health tech field includes:

- Technology in support of active and healthy living and aging
- Digital technologies for health- and social-care improvement
- Diagnostic, screening, and tracking tools
- Assistive and rehabilitative technologies

The series includes a subseries of books published in partnership with Canada's AGE-WELL that specifically addresses their 8 AgeTech Challenge Areas. Each lecture introduces the context in which the technology is used—wellness, health, medicine, special needs, or other contexts. Authors present and explain the technology and review promising applications and opportunities as well as limitations and challenges. They include material on their own work while surveying the broader landscape of related research, development, and impact.

Donald Juzwishin · Maggie MacNeil ·
Axel Meisen · Paul Stolee

AgeTech Innovations in Healthcare for Older Adults

AGEWELL

Springer

Donald Juzwishin
University of Victoria
Victoria, BC, Canada

Maggie MacNeil
McMaster University
Hamilton, ON, Canada

Axel Meisen
The University of British Columbia
Vancouver, BC, Canada

Paul Stolee
University of Waterloo
Waterloo, ON, Canada

ISSN 2771-7054 ISSN 2771-7070 (electronic)
Synthesis Lectures on Technology and Health
ISBN 978-3-031-86884-9 ISBN 978-3-031-86885-6 (eBook)
https://doi.org/10.1007/978-3-031-86885-6

This Springer imprint is published by the registered company Springer Nature Switzerland AG
The registered company address is: Gewerbestrasse 11, 6330 Cham, Switzerland

This book is dedicated to the memory of those lost in solitude to COVID-19—your strength endures, your spirits remain unbroken, and your stories forever echo in our hearts.

We also wish to dedicate the book to Dr. Janet Fast a pioneering researcher, dedicated educator, and tireless advocate for family caregivers. Her groundbreaking work on the economics of aging and care work has left an indelible mark on policy and practice, improving the lives of countless families across Canada and beyond. May her legacy continue to inspire and guide future generations in the field of family and consumer economics.

Preface

> AgeTech is a term that has come into common parlance during the 2020s. This is due to an increasing interest in identifying ways that technology can play a role in supporting and enabling the health, autonomy, and well-being of older adults whatever their abilities and situations (Sixsmith et al., 2017).

This book is written by a Canadian team. The book, however, takes a global perspective. Canada is a particularly interesting case example because its healthcare systems are recalcitrant and resistant to reforms. Power and politics in the Canadian context combine to sustain long-standing and pernicious problems that are resistant to reforms that would address citizens' unmet health needs. Decades of Royal Commissions, Task Forces and Blue-Ribbon Reports with recommendations with amenable legal, regulatory, policy and program delivery solutions fall on deaf ears when it comes to disturbing the status quo. For this reason, we take a learning health systems' perspective focusing on the unmet health needs of older persons identifying how reforms can be achieved by working in partnership with older persons, healthcare providers, innovators, researchers, and leaders. We draw lessons from other countries that have successfully implemented programs to improve the quality of life of older persons.

Unlike other books in this series, this book focuses on the structures, processes and relationships among the sectors and actors comprising the healthcare services and delivery systems. Many of the challenges and barriers associated with the lack of reform, and noticeably the diffusion of AgeTech, is the *path dependency* in how programs and services are organized and funded. Sustaining a dysfunctional *status quo* becomes the enemy of innovation and, in turn, AgeTech. As we move into an increasingly digital age, characterized recently by the accelerated adoption of generative artificial intelligence, traditional program delivery of care and support of older people becomes more outmoded and less fit for purpose.

We are under no illusion that AgeTech is a magic bullet in the successful reform of healthcare delivery systems. Those reforms must be catalyzed through democratic legislative, regulatory, policy, funding, and program delivery initiatives. We provide evidence that AgeTech can be an important ingredient in helping achieve a continuous and effectively integrated healthcare delivery system committed to continuous learning. This book focuses on digital technologies, while other books in the Springer AGE-WELL series address specific challenges and issues in greater detail:

- supportive homes and communities
- autonomy and independence
- mobility and transportation
- cognitive health and dementia
- financial wellness and employment
- staying connected
- healthy lifestyles and wellness.

The book is centered on older persons, their needs, aspirations, and challenges making the audience for this book wide ranging. Older persons are surrounded by a multitude of providers many of whom are not aware of one another's role or responsibility. Without a common understanding and alignment of all providers in their respective roles, achieving a high performing healthcare delivery system can be very challenging.

The book is targeted at a multifaceted audience. It will be of interest to leaders in large, medium, and small enterprises that are committed to understanding and addressing the complexities of partnering with health system providers, funders, and procurement. Educators and researchers in postsecondary institutions in fields of health care will welcome the ideas and recommendations put forth. Innovators who often find themselves working in isolation and are therefore distant from older adults will want to understand and work closer with those they are attempting to help. Finally, the book is targeted at citizens, legislators, regulators, policymakers, and providers to supply them with the evidence that will inspire them to mobilize reforms necessary to improve health services delivery and the quality-of-life of older adults.

Victoria, Canada — Donald Juzwishin
Hamilton, Canada — Maggie MacNeil
Vancouver, Canada — Axel Meisen
Waterloo, Canada — Paul Stolee

Reference

Sixsmith, A., Mihailidis, A., & Simeonov, D. (2017). Aging and technology: Taking the research into the real world. *Public Policy & Aging Report*, *27*(2), 74–78. https://doi.org/10.1093/ppar/prx007

Acknowledgements

The authors would like to thank the Science and Technology for Aging Research (STAR) Institute, of Simon Fraser University, for its administrative and financial support, and AGE-WELL for its financial support. We wish to acknowledge the fastidious support of J. Lynn Fraser for the glossary, indexing, and copy editing of the book. A special thank you to Juliet Neun-Hornick at Simon Fraser University for her administrative and coordinating roles with our team and with Springer.

The authors would particularly like to thank everyone in the AGE-WELL community—Network Management Office, researchers, older adults and caregivers, and partners. Their passion for innovation was essential in developing and sustaining this book's creation. Thank you for your collaborative spirit.

AGE-WELL (www.agewell-nce.ca) is Canada's Technology and Aging Network. The pan-Canadian network brings together researchers, older adults, caregivers, partner organizations, and future leaders to accelerate the delivery of technology-based solutions that make a meaningful difference in the lives of Canadians. AGE-WELL researchers are producing technologies, services, policies, and practices that improve quality of life for older adults and caregivers and generate social and economic benefits for Canada. AGE-WELL's work is supported through Government of Canada funding programs.

The Science and Technology for Aging Research (STAR) Institute (www.sfu.ca/starinstitute) at Simon Fraser University (SFU) is committed to supporting community-engaged research in the rapidly growing area of technology and aging. The Institute supports the development and implementation of technologies to address many of the health challenges encountered in old age, as well as addresses the social, commercial, and policy aspects of using and accessing technologies. STAR also supports the AGE-WELL network.

AGEWELL

STAR INSTITUTE
Science and Technology for Aging Research
Simon Fraser University

Competing Interests The authors have no competing interests to declare.

Contents

Part I Part One

1 Introduction ... 3
The Challenge ... 3
AgeTech Defined ... 3
Health Care Services Organization, Policy and Functioning ... 4
The Origins of AgeTech ... 5
The Business of AgeTech ... 6
What This Book Is About ... 6
The Objectives of the Book ... 7
Approach and Outline of the Book ... 7
References ... 9

2 Health and Social Needs of Older People in the Twenty-First Century ... 11
The Challenge ... 11
Persona and Scenario—Juanita and Rodrigo ... 11
The Five Senses ... 12
Sight ... 12
Hearing ... 13
Touch ... 13
Taste ... 14
Smell ... 14
Cognition ... 14
Skeletomuscular Functions ... 15
Global Challenges of Aging, Chronic Diseases, Co-morbidity and Multimorbidity ... 15
The Challenges for Future Older Adults and Governments ... 17
Research Priorities ... 17

Bringing Research and Innovation Closer to Home and Focusing on the Older Adult ... 18
Aging and Multimorbidity ... 18
Chronic Diseases ... 19
Networks, Collaboration and Team Care ... 20
Perspectives of Older Adults with Multiple Chronic Conditions Compared to Their Caregivers and Health Care Providers ... 20
Concluding Remarks ... 22
References ... 22

3 Aging at the Nexus of AgeTech and the Evolution of Health Care and Health Service Delivery Reform ... 25
The Challenge ... 25
Persona and Scenario—Margaret ... 25
Ageism ... 26
High-Performing Health Care Systems ... 27
Historical Development and Evolution ... 28
First Principles of AgeTech in Health Care Delivery ... 30
Health Care Policymaking ... 31
Life Cycle Health Technology Assessment ... 33
Why Reform Is So Recalcitrant and How Can AgeTech Help? ... 34
Diffusion of Innovations in Health Service Organizations ... 34
Knowledge, Innovation and Impact ... 35
How Can AgeTech Come to Margaret's Aid? ... 35
Preparing for the Future ... 36
Concluding Remarks ... 37
References ... 37

Part II Part Two

4 Health Care and Health Service Delivery for the Home Environment ... 41
The Challenge ... 41
Persona and Scenario—Brian and Clea ... 41
AgeTech and Home Care ... 44
Examples of Improved and New Technologies and Systems ... 46
Touch Activated Human–Machine Interfaces for Care Recipients ... 46
Other Human–Machine Interfaces and Virtual Digital Assistants for Care Recipients and Caregivers ... 48
Magnification and Other Visualization Technologies ... 49
Mechanical Assistive Technologies ... 50
System Technologies for Remote Care Monitoring and Provision ... 51

Anticipatory Digital Technology Systems ... 52
Comprehensive Systems ... 52
Seminal Demonstration Project ... 53
Concluding Remarks ... 54
References ... 55

5 **Long Term Residential Care** ... 57
The Challenge ... 57
Persona and Scenario—Agnes ... 57
Long Term Residential Care ... 59
The Ethical and Financial Imperative ... 64
AgeTech in Long Term Care Homes ... 64
Using Technology to Support Person-Centered Care ... 66
Technologies that Could Help Agnes and Other LTC Residents ... 67
Health Care Information and Management Technologies ... 67
How Could These Technologies Help Agnes? ... 70
Technologies to Support Communication and Social Interaction ... 70
How Could These Technologies Help Agnes? ... 71
Technologies for Entertainment and Recreation ... 71
Social Robots and Automated Prompting Systems ... 73
How Could These Technologies Help Agnes? ... 76
Remote Monitoring Technology, Including Sensors and Wearables ... 76
How Could These Technologies Help Agnes? ... 77
Seminal Demonstration Project ... 77
Concluding Remarks ... 78
References ... 78

6 **Rehabilitation** ... 91
The Challenge ... 91
Persona and Scenario—Jin ... 92
How Can Rehabilitation Help Older Adults? ... 93
AgeTech and Rehabilitation ... 95
Robotics and Rehabilitation ... 98
Virtual Reality ... 101
How Could These Technologies Help Jin? ... 104
Lessons from the COVID-19 Pandemic ... 105
Engaging Patients in AgeTech for Rehabilitation ... 106
Opportunities: AgeTech Training for Rehabilitation Patients and Providers ... 106
AgeTech Training for Older People and Their Families ... 107
Low-cost AgeTech for Rehabilitation ... 108

Seminal Demonstration Project 108
Concluding Remarks 109
Find Out More 109
References 110

7 **Primary Care and Care Coordination** 119
The Challenge 119
Persona and Scenario—Mariam 119
Background 120
Opportunities for Technology in Primary Care and Care Coordination 124
Virtual Care 125
How Could These Technologies Help Mariam? 127
Mhealth and E-visits 128
How Could These Technologies Help Mariam? 129
Technology-Enabled Primary Care and Care Coordination 130
Seminal Demonstration Project 130
Concluding Remarks 131
References 131

Part III Part Three

8 **Electronic Health Records** 139
The Challenge 139
Persona and Scenario—Dominic 139
EHR Use for Clinical Practice and Patient Safety 141
Online Patient Portals 142
Embedded Risk and Frailty Screening 143
Potential Benefit of EHRs Unrealized 143
Factors Affecting EHR Adoption 144
Promising Initiatives 145
How Could These Technologies Help Dominic? 145
Seminal Demonstration Project 146
References 146

9 **Aging and Falling Through the Cracks of Health Care Delivery** 151
The Challenge 151
Persona and Scenario—Francisco 152
Who Are the Older Adults Living in Poverty? 153
Where and What Is the Interstitial Space? 154
Technologies to Bridge the Gap 155
The Ethical and Practical Imperatives 155
Ageism 156

Access to Service 157
Equitable Distribution of Health 159
Privacy and Confidentiality 160
Building Trust 162
Diminishment of Power and Influence 163
Autonomy, Safety and Security 164
Quality of Care and Value for Money 164
Rubric for Advancing AgeTech to Address Unmet Needs 166
Seminal Demonstration Project 167
Concluding Remarks 167
References 169

10 A Vision for Health Care and AgeTech in the Twenty-First Century 171
Introduction 171
The Role and Importance of Reforms in the Legislative and Policy Landscapes 171
An Idealized Health Care System 174
The Life Trajectory Approach 174
AgeTech, Aging and Values 176
Concluding Remarks 178
References 179

11 Concluding Remarks About the Future 181
Introduction 181
Addressing System Dysfunctionality Through Reform Through Research and Innovation 183
The Tragedy of the Siloed Health Service Mentality 183
Lessons of the COVID-19 Pandemic 184
AgeTech a Global Phenomenon 185
AgeTech Is Not a 'Silver Bullet' But It Can Be a Catalyst for Reform 185
AgeTech and Self-care 185
Co-creation of AgeTech 186
Co-designing AgeTech 187
Ageism 188
Concluding Remarks 189
References 189

Appendix A: Questions and Challenges for Readers 193

Glossary 201

Index 223

Acronyms

AAC	Augmentative and alternative communication
ADL	Activities of daily living
AHRQ	Agency for Healthcare Research and Quality
AI	Artificial intelligence
APPTA	Advancing Policies and Practices in Technology and Aging
AT	Assistive Technology
AUA	Assessment Urgency Algorithm
CABHI	Centre for Aging + Brain Health Innovation
CADTH	Canada's Agency for Drugs in Technology and Health
CAPs	Clinical assessment protocols
CDHE	Centre for Digital Health Evaluation
CIHI	Canadian Institute of Health Information
CIHR	Canadian Institutes of Health Research
CIRCA	Computer Interactive Reminiscence and Conversation Aid
COVID-19	Coronavirus disease
CPOE	Computerized provider order entry
CTAAN	Centre of Technology Adoption for Aging in the North
D4D	Design for Dignity
DiME	Digital Medicine Society
ECG	Electrocardiogram
EHRs	Electronic health records
EMR	Electronic medical record
ePRO	Electronic patient reported outcomes
EU	European Union
FI	Frailty Index
FNHA	First Nations Health Authority
GP	General Physician
GPS	Global positioning system

H3	Healthcare and Health service delivery for the Home environment
HSO	Health Standards Organization
HTA	Health technology assessment
ICT	Information communication technology
IFR	International Federation of Robotics
interRAI	Inter residential assessment instrument
interRAI-LTCF	Inter residential assessment instrument—long term care facility
IT	Information technology
LC-HTA	Life cycle—health technology assessment
LTC	Long-term care
LTCF	Long-term care facility
MCC	Multiple chronic conditions
MDS	Minimum Data Set
NHS	National Health Service
OECD	Organisation for Economic Co-operation and Development
PRN	Pro re nata (as things stand)
PSW	Personal service workers
QR code	Quick response code
RAI-HC	Resident assessment instrument—home care
RFID	Radio frequency identification
ROMA	Research-oriented managed access
SDOH	Social determinants of health
SELMA	Self-management of chronic pain
SOS	Shiny object syndrome
SPEC	Systems for Person-centered Elder Care
STAR	Science and Technology for Aging Research
U.N.	United Nations
U.S.A.	United States of America
UNDESA	United Nations Department of Economic and Social Affairs, Population Division
UNESCO	United Nations Educational, Scientific and Cultural Organization
USD	United States dollar
VR	Virtual reality
W21C	Ward of the 21st Century—places of health care research and innovation in Calgary (Canada)
WHO	World Health Organization
Wi-Fi	Wireless network technology

Numbers

2D	Two dimensions
3D	Three dimensions
4Cs	First contact, comprehensiveness, coordination and continuity

Part I
Part One

Introduction 1

The Challenge

The United Nations (UN) declaration of the UN Decade of Healthy Aging (2021–2030) (World Health Organization, 2023) coupled with the declaration in March 2020 by the World Health Organization (WHO) of a global COVID-19 pandemic signaled, irrevocably, the way health care would be delivered during the pandemic and in the foreseeable future. AgeTech was no longer a *nice to have* or a discretionary accoutrement in the tool kit of health care services. It was, instead, mobilized as an essential means for citizens to gain access to care and communicate with family, health care providers and institutions. The COVID-19 pandemic also brought into glaring focus a multitude of weaknesses and challenges with health care delivery systems that existed before the pandemic started. Older adults were more severely impacted by the pandemic than any other age group (Fig. 1.1).

AgeTech Defined

AgeTech, for the purpose of this book, is defined as any digital technology that aims to disrupt, extend, expand or re-organize existing health and social practices, leading to new, more agile opportunities to enhance independent living of older adults in comfort and security. AgeTech is an important and necessary ingredient for the health care delivery system to provide services to improve the lives of older adults especially those with physical or cognitive challenges.

This book will identify the weaknesses in conventional health care delivery systems with a view to identifying legislation, regulations, policies, funding models and program delivery reforms that when coupled with AgeTech could ameliorate the challenges. The

D. Juzwishin et al., *AgeTech Innovations in Healthcare for Older Adults*, Synthesis Lectures on Technology and Health, https://doi.org/10.1007/978-3-031-86885-6_1

Fig. 1.1 Together in care. *Credit* Shutterstock: Worawee Meepian, https://www.shutterstock.com/image-photo/mature-female-elderly-care-facility-gets-1805972035

book will prepare the reader to understand co-development and co-production to work with older persons and the health care delivery systems to catalyze the necessary improvements that will optimize benefits through the effective use of AgeTech ultimately resulting in improved health states for older adults.

Health Care Services Organization, Policy and Functioning

Health care services are one of the most complex undertakings in a modern society. How should programs and services to older adults be delivered? How do service delivery and AgeTech interact? Are they coordinated? Are there conflicting objectives between and among enabling legislation, regulations, funding models, policies, providers and their practices? Is AgeTech meeting older adults' needs and wants? Having the necessary ingredients for an effective health care services and delivery system is important *but if their interactions are dysfunctional and or they are unable to coordinate, collaborate, integrate and communicate among one another* the health care necessary at the time and location that older persons need will have failed. This book will help readers ensure that they understand and tackle the root of the problem.

Health care services and their delivery are oriented toward improving and sustaining the health of its citizens through programs of health promotion, primary disease prevention, primary care, medical treatment, rehabilitation and home care coupled with social support. Maximizing the health outcomes of citizens so that they can enjoy a high quality-of-life is a primary objective of any society supported by their health care service and delivery systems.

AgeTech emerges from the convergence of digital technologies and other technologies that aim to serve an aging population. Today's health care services and delivery systems are not well prepared or equipped to identify, co-create, adopt, adapt, assess or implement the opportunities offered by AgeTech because of legislative, regulatory, policy, funding, practice and program characteristics from the past that serve as barriers to today's innovation. We attribute this inertia within the health care delivery system to *path dependency*.

Health care systems are established on a foundation of legislation, regulation and programing that are in response to yesterday's issues through a predictable pattern of behavior and expenditure that attempts to maximize the outcomes for citizens from the interventions. The accelerated emergence of innovative digital technologies has outpaced the ability of health care delivery systems to take advantage of them. Our goal is to help readers understand the policy environment and to prepare them to work effectively with older adults, caregivers, providers, researchers, innovators, and industry to stimulate reforms that improve health service delivery to older adults.

The Origins of AgeTech

AgeTech is a term preceded by the terms Gerontechnology, Age Tech, and age tech in reference to research and principles in the field of technology and aging. The term Gerontechnology originated with Graafmans and his team at the Eindhoven University of Technology where a social worker requested engineers be involved in responding to problems encountered by older adults, thus the term was coined in 1988. For a full description of the further development of gerontechnology, translating the needs of older adults to the engineering community and the structuring of the research agenda the reader is referred to Sunkyo's book on gerontechnology (Graafmans, 2016). The word is a tongue twister and so the shorthand version of the field became Age Tech and in Canada the AGE-WELL network adopted the term AgeTech.

The Business of AgeTech

Since the emergence of AgeTech as a field of research and practice in 1988, entrepreneurial developments have escalated rapidly. Etkin released a book in 2022 titled *The AgeTech revolution: A book about the intersection of aging and technology* (Etkin, 2022) in which she describes the digital transformation taking place in our societies and how they can have a positive impact on the lives of older adults. Etkin describes cases of entrepreneurs, engineers, and technologists tackling the challenges of merging the business opportunities that aging and technology provide leading to the improvement of the lives of older adults. The challenge, however, in many health care systems is the recalcitrant nature of adoption and diffusion of new and emerging effective technologies in health care delivery. Etkin on her website, *The Gerontechnologist*™ (Etkin, 2023), illustrates the many entrepreneurial initiatives that are taking place around the globe. How than do we find ways to improve the linkage and exchange among the research, practice and business communities to provide older adults with the best AgeTech?

What This Book Is About

In this book we identify the issues and barriers to AgeTech diffusing effectively. We attempt to prepare the reader to develop and implement responsible and accountable risk mitigation strategies, techniques and tools that will stimulate the safety and effectiveness of health care delivery while innovating with agile approaches. AgeTech is not a silver bullet nor is it a simple solution to the complex, nuanced, multifaceted and often recalcitrant challenges facing us. AgeTech is one tool, which if mobilized effectively with necessary health reforms, can help address many of the issues older adults face when interacting with the health delivery systems.

The increasing rate of diffusion of digital technologies in the past thirty years has been unprecedented in societies' adoption of computers, smart portable devices, and the use of the Internet (McGrath, 2019). The convergence of digital technologies with other household, institutional and community technologies including automobiles, cameras, cardiac arrythmia monitors, closed loop glucose monitors and motion sensors, have created an opportunity for safer, convenient and more comfortable living environments for older citizens in our society. This book provides readers with concepts, tools, approaches, examples and guides for how to increase the chances that AgeTech will optimize the quality-of-life of older adults.

The Objectives of the Book

The objectives of this book are to provide the reader with an understanding of health care systems and their components with a view to arming them with approaches and tools that will advance the use of AgeTech to improve the quality-of-life of older adults:

- We review and identify the challenges and opportunities in the biological and social processes of aging using personas and scenarios within the personal, social, and clinical contexts and how these challenge current and future health care delivery systems and practices.
- We review, at a global level, the emergence and development of health care systems providing the reader with an appreciation of the barriers and opportunities of health care reform within the context of policymaking, AgeTech and the life cycle of health technology assessment.
- We focus on and examine the challenges and opportunities in home care, rehabilitation care, long term care (LTC), primary care with a view to providing the reader with the knowledge, skills and tools to help improve the lives of older citizens.
- We identify and explore the issues surrounding the poor adoption of a standardized electronic health record and its key importance as the citizens record of care and its accessibility and interoperability within the health care delivery system.
- We introduce the reader to a vision, conceptual models and ethical values that describe and explain the challenges and issues associated with health care services and identify opportunities for future improvement.
- We provide readers with examples of AgeTech from research, innovations, pilot projects and other promising initiatives that respond to the health care needs of older citizens.
- We provide readers with approaches, tools and methods that will help expedite the development, implementation, and evaluation of effective AgeTech.
- We identify seminal projects and areas for future research and innovation.
- We provide a list of questions and issues that readers can ask of themselves to continue stimulating thinking about advancing reforms coupled with AgeTech.

Approach and Outline of the Book

The book is comprised of three parts divided into eleven chapters.

Part one of the book begins with Chap. 2 providing a brief global description of the health and social needs and challenges associated with aging. This will lead to identifying the challenges that health care systems face today and will need to be addressed in the future. Chapter 3 provides an orientation to health care delivery systems and services with an explanation of policymaking, the first principles of AgeTech, the life cycle of health technology assessment and an explanation of organizational conditions and characteristics that must be considered to facilitate effective diffusion of innovations. The book sets the stage for addressing contemporary challenges of older adults and the problematic convergence of aging and AgeTech as a global phenomenon. An explanation of why health care reform is so hard to achieve is provided.

Part two starts with Chap. 4 focusing on the challenges and opportunities for AgeTech and health care delivery in home settings. Chapter 5 examines the challenges and opportunities for AgeTech in long term residential care. Chapter 6 examines the application of AgeTech in rehabilitation or restorative program areas. Chapter 7 explores the challenges and issues associated with AgeTech, primary care and the coordination of services.

Part three of the book looks toward the future. Chapter 8 introduces us to the challenges posed by the lack of a standardized EHR and the need for its effective interoperability within the health care delivery system. Chapter 9 examines the challenges facing older citizens that fall through the cracks in the health care system and how coupling policy reforms and AgeTech could help ameliorate them. Chapter 10 introduces and develops several conceptual models that bring the structures, processes, values and outcomes of healthcare systems to life catalyzed by the effective utilization of AgeTech. Chapter 11, the concluding chapter, identifies and addresses emerging methodological, ethical, policy, knowledge translation, implementation and technology assessment issues along with several limitations that confront the advancement of AgeTech in the health care delivery systems. Finally, the chapter identifies some promising future directions for addressing these issues.

Throughout the book challenges facing the health sector are described by personas and scenarios that highlight one or several issues confronting that type or level of health care delivery. An overview of AgeTech in this type of health care setting is provided with the identification of where the cracks and challenges exist. Highlight boxes are used intermittently to identify promising AgeTech or research projects that have been completed or are underway. Technology products and services coupled with key initiatives demonstrate how an innovation can be mobilized is provided. The reader is introduced to approaches, means and tools by which researchers and innovators have co-created solutions with older adults, small- and medium-enterprises, and large enterprises demonstrating how they can be leveraged and scaled up to benefit more citizens. Ideas and undertakings in the form of seminal projects are described to inspire further *research* and *innovation*. The book's final chapters offer concluding remarks along with a reading list for those readers who wish to access other resources.

Appendix A at the end of the book presents readers with a series of questions and issues that the reader is encouraged to think about and explore innovative solutions in the context of legislative, regulatory, policy and funding contours of the health care delivery landscape.

References

Etkin, K. (2022). *The AgeTech revolution: A book about the intersection of aging and technology.* New Degree Press.

Etkin, K. (2023). *The Gerontechnologist^{TM}*. https://thegerontechnologist.com/

Graafmans, J. A. M. (2016). The history and incubation of gerontechnology. In S. Kwon (Ed.), *Gerontechnology: Research, practice, and principles in the field of technology and aging* (2nd Ed., Chap. 1). Springer Publishing. https://connect.springerpub.com/content/book/978-0-8261-2889-8/part/part01/chapter/ch01

McGrath, R. (2019, September 25). The pace of technology adoption is speeding up. *Harvard Business Review.* https://hbr.org/2013/11/the-pace-of-technology-adoption-is-speeding-up

World Health Organization (WHO). (2023). *WHO's work on the UN decade of healthy ageing (2021–2030).* https://www.who.int/initiatives/decade-of-healthy-ageing

2 Health and Social Needs of Older People in the Twenty-First Century

The Challenge

The health and social challenges and issues for older people that COVID-19 brought into focus have highlighted unmet needs as well as opportunities for societies to plan and prepare for more effective and appropriate health and social service supports in the future. In this chapter we identify the pressures and challenges the future holds for aging older persons. This understanding will inform strategies and plans that may improve the agency and quality-of-life of older persons in the future as well as reforms of health systems and services (Fig. 2.1).

We will begin our overview of the biological and physiological characteristics of aging by focusing on the five senses and the related assistive devices that have compensated for a loss of those senses.

Persona and Scenario—Juanita and Rodrigo

Juanita and Rodrigo in their mid 70s have been married for 50 years. They live in an apartment in a small town of 2000 inhabitants. There is a primary care clinic in their community as well as a small hospital and basic home care services. The nearest large center is 100 km away. They have a close circle of friends and five grown children who are away from home to pursue their education and employment. The couple live on a modest fixed income but as they age, they are finding that aging brings with it a loss of sight, hearing and cognitive abilities. Rodrigo is beginning to show symptoms of early-stage dementia that is compounded by Type II diabetes which he is attempting to control through diet and exercise. Rodrigo has lost his driver's license and Juanita does not drive.

D. Juzwishin et al., *AgeTech Innovations in Healthcare for Older Adults*, Synthesis Lectures on Technology and Health, https://doi.org/10.1007/978-3-031-86885-6_2

Fig. 2.1. Navigating life together. *Credit* Shutterstock: fizkes, https://www.shutterstock.com/image-photo/positive-young-indian-doctor-woman-showing-2389637553

Juanita has a very active social life and is involved in several community projects supporting local neighbors who themselves are dealing with health and mobility issues. Juanita is beginning to show early signs of macular degeneration.

The Five Senses

The five senses and their related assistive devices are essential enablers of quality human life. An emerging field of promise are the application of virtual reality coupled with generative artificial intelligence to stimulate and enhance the senses.

Sight

The loss of sight is often a gradual occurrence and over the years simple technology has been developed to compensate for this loss. Sight improvements have mainly resulted from improved physical devices, including better eyeglasses, contact lenses and artificial lenses, many of which are currently produced using digital technologies, especially numerical machining. The devices themselves generally do not have significant digital content. Digitally enabled sight and sight enhancement are still in the realm of early research and development. An exception is magnification that enables, for example, reading fine print on medicinal and food packaging. Such magnification is well within the capabilities of current and next-generation mobile phones. These opportunities are further addressed in subsequent chapters.

Hearing

In contrast to assistive sight devices, digitally enhanced hearing aids are now widely available at a reasonable cost. Their advances were facilitated by the miniaturization of digital circuits, microphones, speakers, power supplies and improvements in manufacturing technologies. However, important deficiencies remain. These include limited battery life, interference and feedback, low water resistance, wearing discomfort and operational difficulties. The deficiencies arise from a combination of design features, especially the small size of devices, and the limited visual acuity and dexterity of many aging users. Inadequate instructional manuals and training are also contributing factors (Bulğurcu et al., 2020).

Touch

The sense of touch includes sensations of position and movement, temperature and pain. Unlike other senses, touch is multidimensional and distributed over the entire human body. When the sense of touch is impaired (i.e., when hypoaesthesia occurs) due to advancing frailty and adverse medical events such as injuries, diabetes and amputations, its complexity makes rehabilitation very difficult. Artificial limbs have seen great advances mechanically, largely enabled by new materials and forming processes, but they typically do not impart a sense of touch comparable to their natural, healthy counterparts. Their digital content is still low, but progress based on digital technologies is occurring such as in the example of a skin-inspired organic digital mechanoreceptor (Tee et al., 2015). Touch also enables temperature and pain sensitivities that are important for safety. Both sensitivities are difficult to restore (with or without digital assistance) because they affect all locations of the body. Finally, the sense of touch provides more than physical stimuli as it can also convey emotions and can be a form of communication.

The complexities of touch, including its historical and philosophical dimensions, have been explored in detail by Prescott et al. (2016) and Paterson (2020). Touch provides a rich field for innovative research that merits pursuit. An interesting example is the skin-inspired digital mechanoreceptor described by Tee et al. (2015). However, most current devices enhancing touch are still far from practical use.

Taste

Taste and smell are also important from a safety perspective, especially with respect to judging the freshness of food and drink. While impaired senses of taste and smell are difficult to restore, digitally enabled devices can detect certain contaminants and provide visual or audio alerts. Examples are the detection of pathogens in milk and *Escherichia coli* (E. coli) in foods. However, most detection methods are only suitable for use in industrial and laboratory settings rather than home environments. Zeinhoma et al. (2018) describe a portable fluorescent imager that functions in conjunction with smartphones, thereby holding promise for future low-cost home use.

Smell

Similarly, there are no well-functioning, commercial digital devices that effectively restore smell, but a wide variety of equipment is available, including small, digitally enhanced instruments that detect chemical compounds that are difficult for the human olfactory system to discern. They include common smoke, carbon monoxide and hydrocarbon detectors as well as more sophisticated equipment used in industrial settings. Ferraro et al. (2019) provide a comprehensive review of available industrial equipment, including wearable equipment, which could be adapted for home use to increase the safety of frail and medically compromised aging citizens.

Cognition

In addition to the five senses an older person's cognition capacity will have a material impact on their quality-of-life.

A practical definition of 'cognition' is "the process by which knowledge and understanding is developed in the mind" (Oxford Dictionary, 2023), with knowledge and understanding resulting from inputs received by one or more of the five senses. This is a highly simplified description of cognition, since it provides neither insights into how the inputs are transmitted nor how they are processed.

A projection of the global future burden of serious health-related suffering found that the condition with highest expected proportional increase will be dementia, expected to have a 264% increase by 2060 (Sleeman et al., 2019).

Good evidence exists (especially for fall injuries (Sejdić, 2020) that changes in frailty and cognition are correlated, each contributing to the cycle of decline (Brigola, 2015; Wallace, 2020). While the relationship between cognitive decline and frailty is still uncertain (see for example the work of Halil et al. (2015)), it is clear that these technologies should be designed and function so that they address frailty and cognition jointly.

Skeletomuscular Functions

The skeletomuscular system, i.e., the system consisting of bones, muscles, cartilage, tendons, ligaments, joints and other connective tissue, supports essential bodily functions of human life. With advancing age, components and the skeletomuscular system weaken, often accelerated by disease, falls and accidents. The nervous system, which controls the skeletomuscular system, also deteriorates.

Comprehensive overviews of the impact of aging on the skeletomuscular functions are given by Fernandez and Teasdale (1996) as well as Magee et al. (2016).

In conclusion, the five senses, cognition and skeletomuscular functions are interdependent. Digital technology that enhances any one of them should not be seen in isolation from the others and should take the needs and abilities of the older person into account. For example, better hearing leads to improved cognition and encourages movement and thus maintains skeletomuscular functions.

Global Challenges of Aging, Chronic Diseases, Co-morbidity and Multimorbidity

We will now turn to a 10,000 m view of the challenges that chronic disease, co-morbidities and multimorbidity's present to an aging global population. To appreciate the full force of the burden of illness of older adults today on the global scale it is instructive to view the United Nations Population Division dynamic population pyramids throughout the 2000's and compare them to the projections to 2100 (2022). The projections have several assumptions built into them that will affect their accuracy. However, one can appreciate that the challenges facing older adults will be magnified many times over if steps are not taken in the intervening years to adjust state health care delivery programs to meet the health needs of older adults.

A comparison of the two graphs illustrates that the world's population of individual's 65 years of age and older are growing more rapidly than the population below that age. The share of global population 65+is projected to rise from 10% in 2022 to 16% in 2050 (United Nations Population Division, 2022). Based on this the United Nations is recommending that "Countries with ageing populations should take steps to adapt public programs to the growing numbers of older persons, including by establishing universal health care and long-term care systems and by improving the sustainability of social security and pension systems" (United Nations, 2022) (Fig. 2.2).

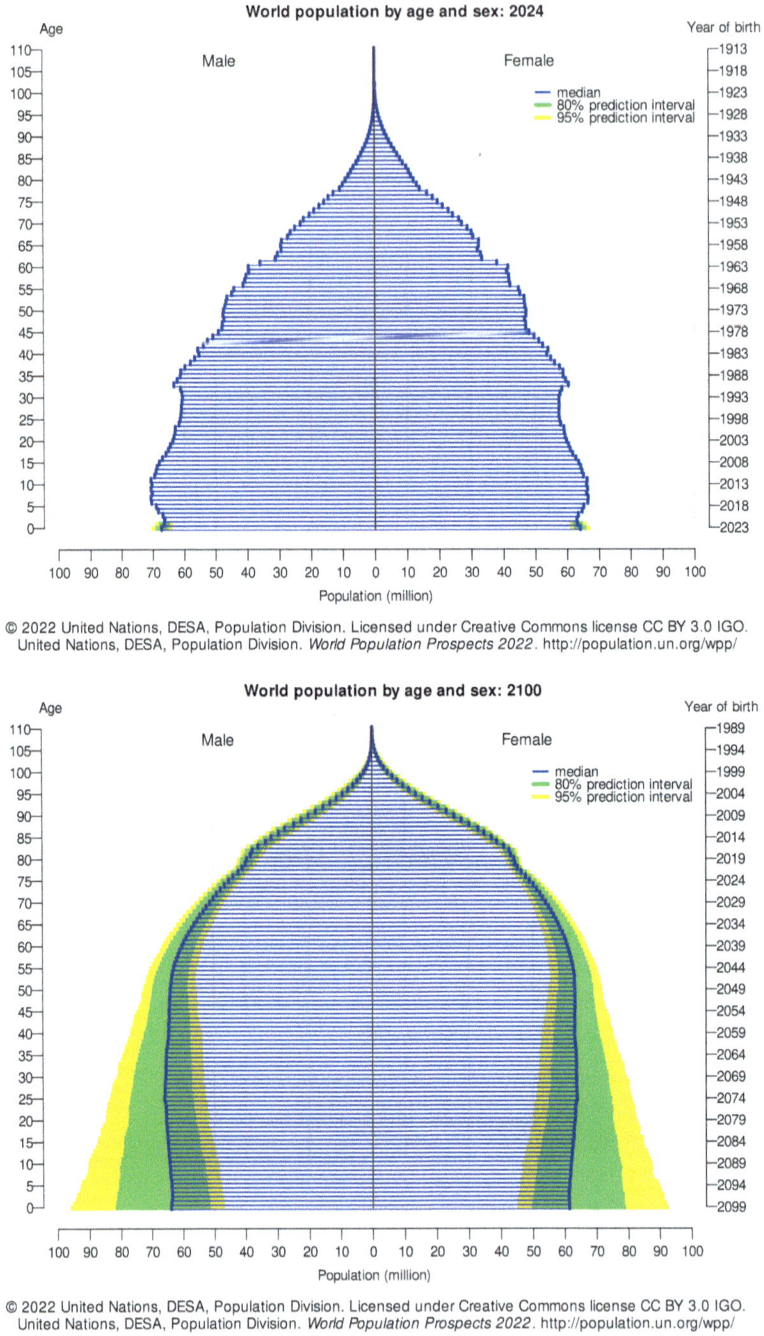

Fig. 2.2 United Nations, DESA, Population Division (2020) from: https://population.un.org/wpp/Graphs/DemographicProfiles/Pyramid/900. Reuse permission granted

The Challenges for Future Older Adults and Governments

With limited resources and strong competition for how those resources should be expended states will need to be innovative in developing strategies, legislation, regulations, policies and program delivery that meet the needs and wants of older adults. Older adults are a heterogenous population. Consequently their care needs and disabilities vary considerably as well. Age is a known risk factor of multiple disease and disability conditions. Aging populations will challenge governments and health care providers to identify and respond to the emerging needs with long term planning and implementation strategies.

Research Priorities

One approach that is proving successful is to convene a wide array of stakeholders in the conversation about how to collaboratively respond to the challenges by placing older adults at the center of the discussion and asking them what they think they need and want. The Canadian Institutes of Health Research (CIHR) convened an Older Adult Advisory Council and Regional Councils to help identify what Canadians thought the strategic directions for future research in Canada ought to be. This resulted in a strategic plan *CIHR Institute of Aging Strategic Plan 2023–2028* (2023) that would reframe aging in Canada where health span, including quality-of-life matches lifespan. The three strategic directions established are:

- Strategic Direction A: Implement Prevention Strategies and Promote Health and Well-being in Aging
- Strategic Direction B: Adapt Health and Care Systems to Better Address the Needs of Older Adults
- Strategic Direction C: Improve Lived and Living Experiences and Quality of Care in the Later Years.

In turn, the research priorities to achieve the strategic directions are to (CIHR, 2023):

Foster a better understanding of the heterogeneity of the older adult population and its impact on health

- Define age in terms of functional ability, intrinsic capacity, environment and biological age, rather than chronological age, to develop a more inclusive understanding of aging.

Mitigate risk related to the development and progression of chronic conditions.

- Support research with a geroscience approach and biological mechanisms of aging.

Focus on wellness and quality-of-life to maintain optimal health and independence.

- Support research on social and structural determinants of health across the life course.

The *CIHR Institute of Aging Strategic Plan 2023–2028*, describes the dynamic interplay among biological age, environment, intrinsic capacity and functional ability. AgeTech can be effective in meeting the wants and needs of older adults if it has a positive influence on the components that impact aging. Leadership at senior government levels will set the stage for regional governments and municipalities to modify their policies and funding patterns along with industry and health care providers to reform their programs to respond to the priorities identified.

Bringing Research and Innovation Closer to Home and Focusing on the Older Adult

We need a better understanding of the relationship of aging and the onset and progression of diseases and how societies can best respond to the challenges these pose. Initiatives like the *CIHR Institute of Aging Strategic Plan 2023–2028* (2023) are essential to provide citizens and governments with the knowledge to introduce legislation, policy and funding to address these challenges. One such approach is AGE-WELL, which is a Canadian network dedicated to the creation of technologies and services that benefit older adults and caregivers. AGE-WELL was launched in 2015 and addresses a wide range of issues in technology and aging through receptor-driven transdisciplinary research, training programs, partnerships, knowledge translation and mobilization as well as the commercial development of technologies (AGE-WELL, 2023).

Aging and Multimorbidity

Multimorbidity affects one in four people (Barnett et al., 2012). It is common in older adults. Globally, the prevalence of multimorbidity is projected to double over the next two decades with population aging (Academy of Medical Sciences, 2018). Nguygen et al. (2020) examined the relationships among multimorbidity,which is generally defined as two or more co-existing chronic conditions, as well as healthy aging and mortality in an English longitudinal study of aging. In the analysis of 9171 older adults aged fifty and over the researchers concluded, "Although it is not impossible for people with multimorbidity to age healthily, those with the most complex combination of diseases are at higher risk of death and have lower levels of health aging" (Nguyen et al., 2020). This finding should stimulate health care providers, innovators and reformers to identify funding opportunities, policies, program delivery and AgeTech approaches that would target supporting higher risk older adults to take preventative approaches preserving or restoring health and improving their quality-of-life thereby reducing their premature mortality.

In their scoping review Maresova et al. (2019), find that chronic diseases of older persons are complex and multifaceted. They determine that these problems require multi-component interventions to avoid the consequences of the chronic diseases. They identify the importance of early detection of the problems that may inevitably lead to disability and restriction of activities of daily activity if they are not "nipped in the bud." This finding is important in underlining the necessity of monitoring and responding to early signals that chronic diseases may lead to further poor health. AgeTech coupled with appropriate health reforms could provide important and timely tools to address the challenges of chronic diseases.

Chronic Diseases

In a profile of Canadian seniors on aging and chronic diseases, Canada's Chief Public Health Officer, Dr. Theresa Tam (Public Health Agency of Canada,2021), reported:

> Four chronic diseases, namely cancer, cardiovascular diseases, diabetes, and chronic respiratory diseases account for over 60% of all deaths in Canada. The onset of these diseases can be delayed or mitigated through changes in behavioral risk factors as well as broader determinants of health.

How can AgeTech help us delay or mitigate the advancement of Juanita and Rodrigo's disease trajectory? One strategy would be to link health care providers more effectively with them to provide them with the tools and knowledge to improve their lifestyle choices. In a scoping review of digital technologies that examined the potential to enable partnerships in the management of chronic diseases Wannheden et al. (2022) identified six features of participatory health technologies that could be helpful. They are (1) support of patient-health provider communication; (2) self-monitoring; (3) tailored self-care support; (4) self-care education; (5) care planning; and (6) opportunities for community peer to peer interactions. The authors concluded that success in using the technologies was dependent on "the importance of clarifying mutual expectations and carefully considering the implications that the introduction of participatory health technologies may have on the work of patients and health care professionals, both individually and in collaboration" (Wannheden et al., 2022). Translating and implementing AgeTech could help address those participatory opportunities between health care providers and older persons.

Networks, Collaboration and Team Care

The importance of health care teams and provision of team care around patients has been recognized within institutional settings like hospitals, LTC and home care for years. However, it has only recently been recognized that a societal wide institutional boundary spanning collaborative team approach is necessary to provide support to aging adults, like Rodrigo and Juanita, in our communities. Caregiver burden, workforce shortages and inequitable access to care exacerbate the challenges. This requires a broader health care community paradigm of programing delivery toward a more collaborative and networked approach in which the respective roles and responsibilities of the partners are clearly articulated and understood. Researchers, innovators, health care providers and aging adults are encouraged to co-develop voluntary forums in which discussions of how best to facilitate and strengthen the collaborative and networked approach can take place. This will require governance, administration and funding entities responsible for the services to develop policies, pathways and guidelines for the effective collaborative exchanges to take place. The unfettered flow of interoperable patient information between and among health care providers is an important ingredient in facilitating a team approach.

AgeTech is an important and necessary tool that can help build the networks, collaborations and continual learning health care providers and organizations.

In response to these challenges the Health Standards Organization (HSO) (2023) has developed and continues to refine health care standards for boundary spanning activities that will improve health, including that of aging adults globally. The HSO does this through three levers of change:

- standards.
- assessment programs.
- education and capacity building in health care organizations.

One of the enablers that the HSO is encouraging health delivery organizations to focus on to achieve an acceleration of reforms is the digital transformation of health care systems.

Perspectives of Older Adults with Multiple Chronic Conditions Compared to Their Caregivers and Health Care Providers

McGilton et al. (2018) undertook a scoping review in which they examined the health and social care needs facing older adults with multiple chronic conditions (MCC). The researchers also examined the perspectives of caregivers and health care providers. The researchers wanted to understand how the social and structural determinants of health impacted those social and health care needs. McGilton et al. (2018) found a confluence of perspectives among older adults, caregivers and health care providers.

McGilton et al. (2018) identified the following unmet needs from all three groups' perspectives:

- Access to information, coordination of services and support.
- Strategies for prevention, maintenance and restoration.
- Training and focus on person-centered approaches.

The policy and program service delivery responses to these important findings must be considered as part of the solution that would expedite health care reforms that improve the flow of information among the actors and their agencies. Digital technologies, AgeTech, are central tools to facilitate the appropriate and essential access as well as flow of information. Putting older adults at the core and allowing them to help define the parameters surrounding the issues of access and flow would help to ameliorate the fragmented state of informatics in our health care systems.

Case Box 2.1 APPTA

APPTA supports governments in generating opportunities to assess the policy and practice challenges of an aging population. This hub is formed through a partnership between AGE-WELL and Research New Brunswick (Research NB). APPTA maintains strong connections with the federal, provincial and territorial governments in Canada by hosting regular one-on-one meetings with policy partners. In 2021–2022, APPTA initiated a new round of engagement with its policy stakeholder network and received 43 submissions for policy questions that required the support of research evidence, a process led by APPTA in consultation with AGE-WELL researchers. APPTA also launched Research Roundup, a new knowledge product. The first issue, released in February 2022, focused on social prescribing (See AGE-WELL/APPTA: https://agewell-nih-appta.ca/our-story/).

At the core of effective communication between caregivers and healthcare providers is the exchange of information about the older adult's care plan, ensuring that any potential safety, health, social or financial consequences are anticipated and addressed. AgeTech can play a crucial role in supporting this process by facilitating tools such as patient journey mapping, video-based exercise instructions, and reminders among other examples.

In respect to structural and social determinants, important factors were:

- Socioeconomic status.
- Cost of transportation, medication, maintaining a healthy lifestyle
- Access to services.
- Coordinating services within and across sectors.
- Educational level.

- Gender and ethnicity
- Isolation especially in women.
- Living in rural settings making access to resource and services difficult.

Concluding Remarks

The authors conclude that reforms, placing the older person's needs and abilities at the center, coupled with appropriate AgeTech that is interoperable among the health and social care systems is required to accommodate the increasing incidence and prevalence of chronic conditions and co-morbidities. Older persons' risk profiles must be monitored, maintained, and updated on a frequent basis. The risk profiles include the older persons' financial capacity, proclivity to isolation and linkages to supports such as food banks or access to transportation.

These reforms complemented by AgeTech can lead to improved integration of care that effectively coordinates the transition between the types and levels of health and social care that is needed by the individual at the right location at the right time with the most appropriate health care provider.

AgeTech can be an invaluable tool to help advance these reforms. One can imagine that every older adult in the community would be registered with a local primary care setting that would be accessible and responsible for care coordination, navigation, making appointments and planning for transportation when the older person is no longer able to. It is essential to protect the confidentiality and security of the older adult, while also enabling them and their family to exercise their autonomy through the available supports.

References

Academy of Medical Sciences. (2018, April). *Multimorbidity: A priority for global health research.* The Academy of Medical Sciences. https://acmedsci.ac.uk/file-download/82222577

Advancing Policies and Practices in Technology and Aging (APPTA). (2021). https://agewell-nih-appta.ca/

AGE-WELL. (2023). *About AGE-WELL.* https://agewell-nce.ca/about-age-well

Barnett, K., Mercer, S. W., Norbury, M., Watt, G., Wyke, S., & Guthrie, B. (2012). Epidemiology of multimorbidity and implications for health care, research, and medical education: A cross-sectional study. *Lancet, 380*, 37–43. https://doi.org/10.1016/S0140-6736(12)60240-2

Brigola, A. G., Rossetti, E. S., Rodrigues dos Santos, B., Neri, A. L., Zazzetta, M. S., Inouye, K., & Pavarini, S. C. I. (2015). Relationship between cognition and frailty in elderly: A systematic review. *Dementia & Neuropsychologia, 9*(2), 110–119. https://doi.org/10.1590/1980-57642015DN92000005

Bulğurcu, S., Irmak, U., Yönem, A., Evren, E., & Çekin, E. (2020). Hearing aid problems in elderly populations. *Ear, Nose & Throat Journal, 99*(5), 323–326. https://doi.org/10.1177/0145561319883526

Canadian Institutes of Health Research. (2023). *Strategic plan 2023–2028: Reframing aging—Empowering older adults [PDF]*. Retrieved from https://cihr-irsc.gc.ca/e/documents/ia_strat_2023_2028-en.pdfcihr-irsc.gc.ca

Canadian Institutes of Health Research (CIHR). (2023, March 24). *CIHR Institute of Aging strategic plan 2023–2028: Reframing aging: Empowering older adults.* https://cihr-irsc.gc.ca/e/53416.html

Ferrandez, A-M., & Teasdale, N. (Eds.). (1996). *Changes in sensory motor behavior in aging* (Issue 114). *Advances in psychology*. Elsevier.

Ferraro, V., Rampino, L., & Stepanovic, M. (2019). A plurisensorial device to support human smell in hazardous environment and prevent respiratory disease, *Advances in Human-Computer Interaction, 2019*, Article no. 3094560. https://doi.org/10.1155/2019/3094560

Halil, M., Kilzilarslanoglu, C. M., Kuyumcu, M. E., Yesil, Y., & Jentoft, A. J. C. (2015). Cognitive aspects of frailty: Mechanisms behind the link between frailty and cognitive impairment. *Journal of Nutrition, Health & Aging, 19*(3), 276–283. https://doi.org/10.1007/s12603-014-0535-z

Health Standards Organization (HSO). (2023). *What is HSO*? https://healthstandards.org/about/

Magee, D. J., Zachazewski, J. E., & Manske, R. C. (Eds.). (2016). *Pathology and intervention in musculoskeletal rehabilitation* (2nd Ed.). Elsevier.

Maresova, P., Javanmardi, E., Barakovic, S., Husic, J. B., Tomsone, S., Krejcar, O., & Kuca, K. (2019). Consequences of chronic diseases and other limitations associated with old age—A scoping review. *BMC Public Health, 19*, Article no. 1431. https://doi.org/10.1186/s12889-019-7762-5

McGilton, K. S., Vellani, S., Yeung, L., Chishtie, J., Commisso, E., Ploeg, J., Andrew, M. K., Ayala, A. P., Gray, M., Morgan, D., Chow, A. F., Parrott, E., Stephens, D., Hale, L., Keatings, M., Walker, J., Wodchis, W. P., Dubé, V., McElhaney, J., & Puts, M. (2018). Identifying and understanding the health and social care needs of older adults with multiple chronic conditions and their caregivers: A scoping review. *BMC Geriatrics, 18*(1), Article no. 231. https://doi.org/10.1186/s12877-018-0925-x

Nguyen, H., Wu, Y.-T., Dregan, A., Vitoratou, S., Chua, K.-C., & Prina, M. (2020). Multimorbidity patterns, all-cause mortality and healthy aging in older English adults: Results from the English longitudinal study of aging. *Geriatrics & Gerontology International, 20*(12), 1126–1132. https://doi.org/10.1111/ggi.14051

Oxford English Dictionary. (2023). *Cognition.* OxfordLearnersDictionaries.com.

Paterson, M. (2020). *The senses of touch: Haptics, affects and technologies.* Routledge.

Prescott, T. J., Ahissar, E., & Izhikevich, E. (Eds.). (2016). *Scholarpedia of touch.* Atlantis Press.

Public Health Agency of Canada. (2021). *Aging and chronic diseases. A profile of Canadian seniors.* https://www.canada.ca/content/dam/hc-sc/documents/services/publications/diseases-and-conditions/aging-chronic-diseases/canadian-seniors-report_2021-eng.pdf

Sejdić, E., Godfrey, A., McIlroy, W., & Montero-Odasso, M. (2020). Engineering human gait and the potential role of wearable sensors to monitor falls. In M. Montero-Odasso & R. Camicioli (Eds.), *Falls and cognition in older persons* (pp. 401–426). Springer Nature Switzerland AG. https://doi.org/10.1007/978-3-030-24233-6_22

Sleeman, K. E., de Brito, M., Etkind, S., Nkhoma, K., Guo, P., Higginson, I. J., Gomes, B., & Harding, R. (2019). The escalating global burden of serious health-related suffering: Projections to 2060 by world regions, age groups, and health conditions. *Lancet Global Health, 7*(7), e883–e892. https://doi.org/10.1016/S2214-109X(19)30172-X

Tee, B.C.-K., Chortos, A., Berndt, A., Nguyen, A. K., Tom, A., McGuire, A., Lin, Z. C., Tien, K., Bae, W.-G., Wang, H., Mei, P., Chou, H.-H., Cui, B., Deisseroth, K., Ng, T. N., & Bao, Z. (2015). A skin-inspired organic digital mechanoreceptor. *Science, 350*(6258), 313–316. https://doi.org/10.1126/science.aaa9306

United Nations Department of Economic and Social Affairs, Population Division. (UNDESA, PD). (2022). *World population prospects 2022: Summary of results*. United Nations. UNDESA/POP/2022/TR/NO. 3. https://www.un.org/development/desa/pd/sites/www.un.org.development.desa.pd/files/wpp2022_summary_of_results.pdf

Wallace, M. K., Theou, O., & Rockwood, K. (2020). Frailty, cognition, and falls. In M. Montero-Odasso & R. Camicioli (Eds.), *Falls and cognition in older persons* (pp. 67–83). Springer Nature Switzerland AG. https://doi.org/10.1007/978-3-030-24233-6_5

Wannheden, C., Åberg-Wennerholm, M., Dahlberg, M., Revenäs, Å., Tolf, S., Eftimovska, E., & Brommels, M. (2022). Digital health technologies enabling partnerships in chronic care management: Scoping review. *Journal of Medical Internet Research, 24*(8), e38980. https://doi.org/10.2196/38980

World Health Organization (WHO). (2015). *World report on ageing and health*. World Health Organization. https://apps.who.int/iris/handle/10665/186463

Zeinhoma, M. M. A., Wang, Y., Song, Y., Zhu, M.-J., Lin, Y., & Du, D. (2018). A portable smartphone device for rapid and sensitive detection of E. coli O157:H7 in yoghurt and egg. *Biosensors and Bioelectronics, 99*(2018), 479–485. https://doi.org/10.1016/j.bios.2017.08.002

Aging at the Nexus of AgeTech and the Evolution of Health Care and Health Service Delivery Reform

3

The Challenge

Older adults around the globe experience vast differences in the quality of their lives. The variation in social, economic, and political forces that result in the health care delivery systems in each country and local regions can vary a great deal. Although there is great variation in the structure and function of health care systems there is a general trend toward providing citizens with universal access to necessary high quality health care. There is significant variation in the degree of universality of access to health care and social supports around the globe. The political scientist Harold Lasswell characterized politics as the interactions among citizens, interest groups, actors and agents of government comprising a society determining who gets what, when and how (Lasswell, 1958). In this chapter we will describe and explain the mechanisms and tools in government that determine what legislation is enacted and what regulations. We will examine the policies that result and that will, in turn, drive the public funding and program delivery of health and social services in a society (Fig. 3.1).

Persona and Scenario—Margaret

Margaret is in her mid 60s and the victim of domestic violence. Her husband Jerry, 10 years her senior, recently passed away from liver cancer and she is now feeling released from the constraints of his violent temper and lack of anger control. Verbally and physically abused Margaret led an isolated life with few friends and acquaintances. She loves her job as a cashier at a local grocery store. Their son and daughter who are independent are estranged from Jerry and Margaret. Margaret is feeling increasingly isolated in her home. However, she realizes that it is important to her to develop new relationships.

D. Juzwishin et al., *AgeTech Innovations in Healthcare for Older Adults*, Synthesis Lectures on Technology and Health, https://doi.org/10.1007/978-3-031-86885-6_3

Fig. 3.1 Bridging distances in care. *Credit* Shutterstock: Nattakorn_Maneerat, https://www.shutterstock.com/image-photo/female-asian-doctor-making-video-call-2159176543

She is now interested in finding meaningful ways in which she can contribute to her community. Because of her age she feels that any knowledge and skills she does have would not be valued by any employers or volunteer agencies. One thought she has is wanting to help other women who are the victims of domestic violence. She also thinks she would like to see more support for victims of domestic violence.

Ageism

Ageism was a term coined in 1968 by Robert Butler, a gerontologist and psychologist, who likened negative stereotyping of older persons to evils of racism and sexism (Bengtson & Whittington, 2014).

In 1991 the United Nations adopted and proclaimed principles for older persons. The principles were adopted because of an appreciation of the diversity in the situation of older persons between countries and between individuals. The increasing proportion of older persons in societies living to an advanced age coupled with an awareness of scientific research dispelling many myths and stereotypes lead to the recognition that older persons should be provided the opportunity to be active contributing members of society.

The proclamation recommended that governments encourage independence, participation, care, self-fulfilment and dignity of older persons.

Governments around the world took on the mantle of improving conditions for older persons at differing rates of progress. Generally, the improvements came in the form of pension reform and social security to provide economic security as well as universal access of all citizens to medical care, hospital care, long-term care and home care as examples. Countries varied in the degree to which they publicly funded access to these services versus what out of pocket costs were expected to be paid by the citizen.

High-Performing Health Care Systems

A systematic review (Ahluwalia et al., 2017) of what defines a high-performing health care delivery system in 2017 found that there was no consistent definition of what characterizes a high performing delivery system or its constituent components. Although there was little agreement on the definition the most frequently identified characteristics included quality, cost, access, equity, patient experience and patient safety. The Commonwealth Fund, which supports research related to health care, has identified five dimensions of performance along with indicators and measures to conduct a comparative analysis of the delivery systems in Australia, Canada, France, Germany, the Netherlands, New Zealand, Norway, Sweden, Switzerland, the United Kingdom and the United States (Schneider et al., 2021). Table 3.1 illustrates the variation among the 11 countries. The United States and Canada rank as the worst performers while Norway, the Netherlands, and Australia rank as the top three. There is much to learn about how different countries are reforming their health care delivery systems to better meet the health needs of their citizens and of interest to us, specifically, are the health needs of older persons.

Table 3.1 Health care system performance rankings

	AUS	CAN	FRA	GER	NETH	NZ	NOR	SWE	SWIZ	UK	US
Overall ranking	3	10	8	5	2	6	1	7	9	4	11
Access to care	8	9	7	3	1	5	2	6	10	4	11
Care process	6	4	10	9	3	1	8	11	7	5	2
Administrative efficiency	2	7	6	9	8	3	1	5	10	4	11
Equity	1	10	7	2	5	9	8	6	3	4	11
Health care outcomes	1	10	6	7	4	8	2	5	3	9	11

Data commonwealth fund analysis

Source Schneider et al. (2021). See https://www.commonwealthfund.org/publications/fund-reports/2021/aug/mirror-mirror-2021-reflecting-poorly

All countries would like to exhibit high performance on the indicators. However, political, economic and Heasocial factors cause significant variation in their capacity to serve their citizens.

In this chapter we provide a brief historical development of health care systems, the first principles of AgeTech, the foundations of policymaking, and why reform is challenging. We also provide a primer on the life cycle of health technology assessment and systematic approaches to advance innovation in health care.

Historical Development and Evolution

Health care and health service delivery systems around the globe are the product of the historical, social, economic and political forces that shape the way these services are structured, organized, funded and delivered. The degree to which services are publicly and or privately funded and delivered differs from country to country as well as between and among regions within a country. Some services are universal, others may have co-payments and yet others will require payment by the client or from insurance programs. The fiscal capacity of countries in supporting their health service delivery can have a wide range. Recognizing this significant variation in the approaches to establishing and funding health service delivery we attempt to frame and identify ways in which AgeTech can be agnostic to the type of funding and delivery in the healthcare system. Our aim is to provide the reader with knowledge and tools to creatively advance the overarching aim of serving older adults to live autonomous and fulfilling lives with the support of AgeTech.

The health service delivery system(s) in a country are generally dictated by legislation, regulations, funding models, policies and health practices. The diffusion of digital technologies is also strongly influenced by them. AgeTech, or digital technologies, can fall into several categories, they can be consumer items, be medically required or they can be both. This creates a challenge for publicly funded health care delivery services to innovate with emerging AgeTech because of the question of who can approve, procure and pay for the promising technology? For example, COVID-19 required quarantine, isolation and social distancing as public health measures to contain the pandemic. Communication between patients and their health care providers often necessitated linkage through the Internet. Not all patients had access to the Internet putting them at a disadvantage of access to health services. This raises the issue of whether some AgeTech that is essential for the effective and efficient delivery of health care services should be provided for by the state. The fiscal capacity of a country, region or organization will limit the degree to which public funding is provided for a universally accessible health service.

Changing values about aging over the last two decades have had a profound effect on how health care services are expected to be delivered to older adults in society. Coupled with this phenomenon is the rapid emergence of AgeTech, which is significant in its ability to support older adults. However, as we will see in the following chapters, because of the complexity and heterogeneity of the aging adult population and its characteristics, it is very difficult to generalize and to assume that one size fits all when it comes to dealing with aging individuals. Recognizing the complexity of health care service delivery as a self-organizing system will help us consider how to drive healthcare reforms, alongside AgeTech, to better address the needs of older adults in society. One approach, central to our prescription, is the use of the social determinants of health (SDOH) to identify and build the bridges between necessary health care reforms and the appropriate use of AgeTech. The other books in the AGE-WELL Springer series provide evidence of where attention must be paid by older adults themselves, as well as policymakers, researchers and innovators to create a society which addresses the aspirational values of older adults.

A new perspective on the health of Canadians, also known as the *Lalonde report*, proposed a reimagined concept of the "health field" (Lalonde, 1974, p. 31). This concept identified four interdependent constituents of society that influence a person's health their biology; environment; lifestyle; and access to health care (Lalonde, 1974, p. 31). Developments in the population health approach to policymaking have resulted in the identification of SDOH. These are the non-medical factors that influence a person's or group's health status. Examples are a secure income, education, food security, housing, social inclusion and access to health services. Health status is often unequally distributed within a society. The WHO states, "This unequal distribution of health-damaging experiences is not in any sense a 'natural' phenomenon but is the result of a toxic combination of poor social policies and programmes, unfair economic arrangements, and bad politics" (Commission on Social Determinants of Health, 2008). Health care delivery, alone, without an effective and responsive public and social care system cannot be successful in addressing the inequitable distribution of health in a population.

Progressive societies have introduced publicly funded universal health care delivery systems over the last 75 years (post-World War Two) to respond to the primary health care needs of citizens through legislative, regulatory, policy, funding, practice and service delivery approaches. Despite progress, there are significant gaps in the health care delivery systems that leave people experiencing poverty at risk.

Health care delivery systems ascribing to the medical model of intervention are often siloed from other SDOH interventions for improving the health outcomes of citizens (White-Chu et al., 2009). The medical model attributes a person's disability to their impairments or differences. Addressing those impairments requires a medical intervention. The social model of disability on the other hand looks at how society is organized and how barriers can be removed to facilitate autonomy of the individual. The way forward is not one or the other approach but a combination of the two. The goal of this book is to identify, describe and inspire the use of digital technologies, data and information (AgeTech) that will add to the dignity, length and quality-of-life of older adults experiencing loss of agency. AgeTech is necessary, but it alone is not sufficient to address the challenges—it will take an "all systems" approach.

The heterogeneous characteristics of the aging adult population and their wide-ranging health care needs have spawned a broad range of services that may or may not be accessible to older adults experiencing poverty through public, private or a mixture of program funding. Programs are allocated funding through global, line-by-line or hybrid budgets, based on historical precedent with adjustments made in response to changes in demand for services. Service delivery, resource consumption (input costs) and outputs are monitored and reported to the jurisdictional funding agency to maintain accountability for the funds expended. In public programs, *working capital* is kept to a minimum and there is very little discretionary opportunity for expenditures outside of the approved program objectives. Publicly funded and administered health care systems are highly programmed in their legal, regulatory and policy responsibilities often leaving them constrained in adapting, adopting and/or financing innovations in response to an unmet need of citizens (MacNeil et al., 2019). Health care organizations often function in 'silos' that are 'blind' to those unmet needs but are satisfied that they are meeting their governance and fiduciary responsibilities. This may achieve the goal of satisfying the auditor general who oversees the prudent expenditure of public funds, but it leaves the older adults with unmet needs!

First Principles of AgeTech in Health Care Delivery

There are several first principles of human phenomena that AgeTech is exposed to in the field of health care delivery. The first is that there is an inherent tension between technology and the human side of health care delivery. This is often referred to as the tension between *high tech* versus *high touch.* The assumption being that the introduction of technology takes away from the humanity of the health care process. An example is a family practitioner sitting at their desk inputting a patient's history into a computer averting their eyes from the patient as they are speaking. Another example would be a nurse practitioner communicating in a virtual care visit with an aging adult in another part of the community without a physical presence. The close contact, touch and empathy may be absent. The challenges are how to bring those important human characteristics into the caring process.

Technology, on one hand, may be a boon in the form of clinically and program effective healthcare interventions, but on the other hand there may be risks of harm that come with that technology that is overhyped or offers false hope. There may be promise of benefit, but it is in fact a false promise. This is referred to as the *hope or hype obsession* (Deyo & Patrick, 2005). An example is an oncological intervention (drug) that through the course of treating the disease also causes pain and suffering to the patient. The role of health care providers, policymakers and regulators are to ensure that the balance of risk or harm is not greater than the benefit accrued.

Another principle is that of the *shiny object syndrome* (SOS). This 'syndrome' posits that humans are in a continual state of distraction brought on by the belief that there is something new emerging or around the corner worth pursuing. An analogy is the aphorism the grass is always greener in the adjacent pasture. In the medical industrial complex where technological innovation and its accelerated pace of diffusion has been remarkable with significant improvements in health outcomes in morbidity and mortality there have also been many failures that have not resulted in improved outcomes. In fact, some have done more harm. For example, thalidomide, which was introduced in the 1950s and 1960s for treatment of nausea in pregnant women, resulted in women giving birth to babies with severe birth defects in thousands of children. A more current example is the introduction of opioids for pain management where there highly addictive nature was not fully understood at the outset. This raised the need for policymakers and providers to base their decisions on validated and robust clinical trials with sufficient power before a healthcare intervention could be introduced to the population. Drugs, biologics and medical devices, many with digital technologies associated with them, require assessment to determine their clinical and cost effectiveness. As we will see in the examples in this book there are many emerging AgeTech that provide significant promise of improving the quality-of-life of older adults, but is the evidence of effectiveness sufficient to convince a consumer or a public health payer or provider to pay for it?

Related to the SOS is *buyer's remorse*, which is the feeling of regret that a consumer experiences after making a purchase or commitment that turns out to be unnecessary, ineffective or inappropriate. This occurs when any technology in the form of a product or service, and digital technology is particularly susceptible, is oversold and does not meet expectations. This phenomenon results in disappointment on the part of the consumer or user and creates an overall feeling of distrust of technology more generally. The remedy to prevent this from arising is an awareness of purchasers, producers and users to realistically describe and manage expectations of the AgeTech to older persons.

Health Care Policymaking

Policy studies, policy analysis and health care policy along with many other policy-focused endeavors emerged from the discipline of political science. Political science is concerned with the systems and dynamics of governance and power that arises from the constitutional and legislative landscape that provides guidance to our governments and societies. As we noted earlier, in Lasswell's terms politics is who gets what, when and how (Lasswell, 1958). The policy process and the making of policy has several important thinkers that we will highlight to support the appreciation of the machinery and complexity associated with the relationship of health care policy and AgeTech. All the models are mere representations of reality, they are not reality. A model or framework is always wrong, but it can help us understand the world better. In an aphorism attributed to Box (1976), "all models are wrong, but some are useful." It is no substitute for the lived experience. The reader is, therefore, cautioned that the model is not meant to be followed in lock step. It is to inform the reader's talents, creativity and imagination in pursuit of a good policy process. Policymaking, therefore, is not a linear path. It is iterative and adaptive to new and emerging conditions.

John W. Kingdon, in his study of public policy, identified three phenomena or streams that must come into confluence for a policy change to be enacted, the problem; the solution; and the political will. A policy window appears when all three come into conjunction (Kingdon, 2011).

For AgeTech to be an accelerant in improving the quality-of-life of older adults the aspired to characteristics of an improved society for older adults must have the synergy and advocacy of citizens in society reaching a consensus that the problem is a priority and needs remedy. Thus policy entrepreneurs and bureaucrats must be convinced and prepared to make the case for reform coupled with political actors who are prepared to exercise the necessary political will to see it materialize.

A second approach to policymaking is described by Howlett et al. (2009) and entails the understanding of the phases of applied problem-solving being matched against the 5 stages of a policy cycle. The five stages are agenda setting, policy formulation, decision-making, policy implementation and policy evaluation. In a civil, democratic and pluralistic society there are many factors that must be accounted for and align for the implementation of policy changes or reforms to address a new set of ideals in society. This is important for AgeTech because it means that many constituents of the society must be in alignment to achieve success in the diffusion and implementation of a new and promising assistive digital technology.

Further insights into policy making are illustrated by the Australian model of policy making which is outlined in the *Australian policy handbook* (Davis & Althaus, 2017).

The cycle goes into further detail identifying the issues surrounding the setting of an agenda, identifying alternatives to address them and exploring policy instruments like legislation, regulations, and funding models that are consulted on with stakeholders. Advocacy needs to be coordinated with other inputs leading to an assessment of alternative policy solutions which may lead to a decision. Each of the stages of the cycle require significant iterative sub cycles within them to reach a consensus for moving ahead. It also means that one may, because of the consultative stage, require a return to the identification of the issues to deal with new insights that arise from the consultation. Confounding this as well are the election cycles that occur every 4 to 5 years in which the elected governments priorities may change the approach to the problem.

The way in which a social problem is defined or framed can a have a significant impact on how it is resolved in the longer term. van der Bijl-Brouwer (2019), in an article on problem framing expertise in public and social innovation, found that framing plays a key role in innovating for complex social problems. Confirming what we identified earlier, public and social innovations are complex and iterative rather than linear. The authors identify that the framing of public and social innovation, such as advancing the opportunities of AgeTech, requires a high level of expertise relying on systemic principles of research, solution generation and reflection driven framing (van der Bijl-Brouwer, 2019). The practical advice for readers of this book is that whether you are a provider, researcher or policymaker your responsibility is to understand and participate in a process to arrive at solutions for the aging adult community in collaboration and to co-create the policy and practice solutions together.

Life Cycle Health Technology Assessment

We identified the importance of the issues of access to services, their quality and costs as being highly determinative of the likely success of innovation and implementation in the health care delivery system. Kirwin et al. (2022) point out that traditional health technology assessment (HTA) processes often involve a significant amount of uncertainty about the clinical or program effectiveness of a health technology (Kirwin et al., 2022). The authors have arrived at a conceptual framework for life cycle health technology assessment that uses evidence appraisal and synthesis with economic evaluation to inform decisions as to what innovations in technology should be supported and those that should be discarded. A technology life cycle consists of a research and development phase where costs are high, income or return is negligible, and prospects of failure are high. If successful, the technology may enter an ascent phase where costs of the previous phase have been recovered and the technology comes to market and diffuses. At maturity use of the technology is stable and is sustained over time. If the utility of the technology is diminished or an alternative arrives it may decline and be rendered obsolete.

Kirwin et al. (2022) points out that HTA processes that align with the managerial and policy making process are more likely to be successful because they reduce analytical time (time in research and evaluation process) and optimize decision uncertainty. In circumstances where there is significant risk to be considered the authors recommend risk-based pricing and research-oriented managed access (ROMA) with real-world data. To achieve the best results for the technology purveyors, funders and providers the authors recommend negotiation of procurement agreements that identify the terms and conditions that are informed by value-of-information methods that draw on routine clinical and administrative data that are linked and coupled with registries. This to reduce uncertainty for both the payer and the innovator. This approach would enable the implementation of high-value high-risk innovations while improving health system sustainability. As the evidence base evolves, the contractually embedded decision rules would simplify the assessment.

With this approach the performance indicators and measures for the broader health care system would be improved by addressing the outcomes for patients, providers, sponsors and payers. Older adults would benefit with earlier access to new and promising technologies, payers would increase the value of technologies they invest in and gain mechanisms to continually review the effectiveness of those technologies over time. Sponsors would benefit because of greater certainty in the outcomes achieved through their investments and expedited access to markets and greater opportunities to demonstrate value.

Why Reform Is So Recalcitrant and How Can AgeTech Help?

The rapidity and sustainability of health care reforms varies significantly among the countries of the world. Countries like the Netherlands and Norway appear to have a highly responsive public policy and reform environment whereas Canada is interminably slow. A former federal health minister, Monique Bégin lamented in an article, "When it comes to moving health care practices forward efficiently, Canada is a country of perpetual pilot projects. We seldom move proven projects into stable, funded programs, and we rarely transfer the outcomes of pilot projects across jurisdictions" (Bégin et al., 2009). Lazar and colleagues (2013) wondered whether the cause was Canada's federal structure wherein provinces have constitutional responsibility for all matters to do with health care and the national government has an overarching responsibility for the *Canada Health Act* that defines the principles of Medicare. The national government leverages its influence through discretionary funding contributing tax points and transfer payments to provinces and territories.

Fierlbeck (2014, p. 478), in her book review of *Paradigm freeze*, summarizes Lazar and colleagues' reasons for the "meagre" progress on health reform due to "specific political interests (such as provincial physician and hospital associations), the fiscal health of a province, the particular political, cultural and institutional history of each region, the relative strength of a governing party, the relationship of each provincial government with its public service, the number (and urgency) of issues competing with health care on the political agenda and, of course, the political will of the government in power explaining why health care reform in Canada has become stymied (pp. 478–480). The way forward is therefore not straightforward and requires a strong awareness of the political landscape in order to be successful.

Bearing in mind the complex dimensions associated with the policy machinery in their country or organization the innovative researcher, health care provider and policymaker will seek to understand what "policy levers to pull" in cooperation and collaboration with other team members and be prepared to act when the "policy window" opens.

Diffusion of Innovations in Health Service Organizations

Having described the challenges and issues of health reform at the macro, national levels, it is also important for the reader to be aware of the social, technical and political dynamics that play out at the mezzo and micro levels of organizations. Toward that end Greenhalgh and colleagues (2005) conducted a systematic review of the conditions and factors that facilitate and act as barriers to innovation in health care. In their work they provide an extensive illustration of the factors and conditions that must be considered and satisfied to facilitate the effective diffusion of innovations.

Readers are encouraged to think about the organization's internal and external environments that they are working within to establish the necessary preconditions for a successful AgeTech innovation to be introduced.

Knowledge, Innovation and Impact

A companion book in this series by Sixsmith et al. (2021) provides guidance to researchers on how to navigate the health care delivery landscape with the view to having the greatest possible impact. The authors point out that having an impact is easy to say but extremely hard to achieve. The book provides practical advice to researchers and innovators on how to work in large teams, engage with stakeholders and clients, how to attract commercialization and facilitate knowledge mobilization as well as how to manage relationships and expectations. A chapter on navigating policy, regulatory and health system landscapes provides practical advice on understanding the messy complex environment and how to work within it. It is imperative to involve end users from the very beginning of a project, ensure a thorough understanding of the regulatory and jurisdictional landscapes, and clearly define the criteria of success, including specific outcomes and measurable indicators of the innovation's performance (MacNeil et al.,2021). The implementation of an AgeTech innovation may also involve an understanding of the budgetary reallocations and trade-offs associated with the AgeTech.

How Can AgeTech Come to Margaret's Aid?

Case Box 3.1

What AgeTech is currently available that Margaret could use to volunteer for a domestic violence hotline?

What avenues of political advocacy would AgeTech provide for Margaret to raise community awareness of the need for support for women of domestic violence?

Preparing for the Future

Recognizing the complexity and recalcitrance of achieving health care reforms to meet older persons' unmet needs the reader is encouraged to develop knowledge and skills on how to navigate and prepare by developing their futures literacy. United Nations Educational, Scientific, and Cultural Organization (UNESCO) defines *futures literacy* as: "the skill that allows people to better understand the role of the future in what they see and do. Being futures literate empowers the imagination, enhances our ability to prepare, recover and invent as changes occur" (UNESCO, 2021). There are many approaches, techniques and tools for thinking about and testing possible futures. Several examples of note are the futures cone (Voros, 2017); guidelines for thinking about the future (Hines & Bishop, 2015); and the Gartner Hype Cycle (Gartner, 2023). Examples of the application of strategic forecasting are provided in Kuosa (2011).

Professor Denise M. Rousseau (2018), from Carnegie Mellon University, recommends employing the science of anticipation to facilitate approaches in organizations that are facing risks and bias associated with decision-making. Rousseau argues for evidence based organizational decisions. Unlike individual decision-makers, organizational decision-makers must work with and through others, deal with political influences that constrain decisions, make many decisions concurrently that affect many stakeholders and may face greater uncertainty due to missing information, a high noise to signal ratio making the interpretation of conditions and change difficult. In addition, there may be limited decision supports and protocols. Rousseau's (2018) recommended approach is to:

- Be careful not to be addressing the wrong problem—interrogate and ask diagnostic questions. Using an incorrect framing of the problem will lead to incorrect solutions.
- Do not ignore the politics—explicitly address the politics of the decision.
- Do not consider just one option—examine and test multiple options.
- Do not focus on a single outcome—identify several outcomes of success and effectiveness.
- Be conscious and guard against narrow interests dominating—broaden the stakeholders considered and involved.
- Rely on easily at hand information—broaden the evidence base to include scientific evidence, data, expert judgment and stakeholder concerns.

These suggestions will resonate well for AgeTech innovators working with organization leaders and policymakers who wish to see their ideas and digital solutions brought into practice.

Concluding Remarks

The lessons from COVID-19 and the accelerated development and diffusion of digital technologies signals that the time is ripe for powerful opportunities in research and development to create an improved world for older adults. The nexus among these factors that we are witnessing is driving new entrepreneurial opportunities, innovations in health service delivery, and a better quality-of-life for older adults and for those who care for them. We are well served to be reminded that technology has a Janus face introducing new opportunities but also a new set of problems. This book focuses on the successful development, assessment and integration of technology as a tool working in partnership with older adults, families, business and industry, researchers and innovators along with governments at all levels to address the key trade-offs necessary to achieve ethically sound health care services that address the unmet needs of older adults.

In part two of this book, we will examine each of the health services sectors to identify what opportunities AgeTech has to offer the improvement in the lives of older adults.

References

Ahluwalia, S. C., Damberg, C. L., Silverman, M., Motala, A., & Shekelle, P. (2017). What defines a high-performing health care delivery system: A systematic review. *The Joint Commission Journal on Quality and Patient Safety, 43*(9), 450–459. https://doi.org/10.1016/j.jcjq.2017.03.010

Bégin, M., Eggertson, L., & Macdonald, N. (2009). A country of perpetual pilot projects. *Canadian Medical Association Journal, 180*(12), 1185. https://doi.org/10.1503/cmaj.090808

Bengtson, V. L., & Whittington, F. J. (2014). From ageism to the longevity revolution: Robert Butler, pioneer. *The Gerontologist, 54*(6), 1064–1069. https://doi.org/10.1093/geront/gnu100

Box, G. E. P. (1976). Science and statistics. *Journal of the American Statistical Association, 71*(356), 791–799. https://doi.org/10.2307/2286841

Commission on Social Determinants of Health. (2008). *Closing the gap in a generation: Health equity through action on the social determinants of health—Final report of the commission on social determinants of health.* World Health Organization. WHO/IER/CSDH/08.1. https://apps.who.int/iris/bitstream/handle/10665/43943/9789241563703_eng.pdf

Davis, G., & Althaus, C. (2017). *The Australian policy handbook: A practical guide to the policy making process.* Routledge. https://doi.org/10.4324/9781003117940

Deyo, R. A., & Patrick, D. L. (2005). *Hope or hype: The obsession with medical advances and the high cost of false promises.* American Management Association.

Fierlbeck, K. (2014). Book review. *Paradigm freeze: Why it is so hard to reform health-care policy in Canada.* H. Lazar, J. Lavis, P-G. Forest, & J. Church (Eds.). *Canadian Public Administration, 57*(3), 478–481. https://doi.org/10.1111/capa.12085

Gartner. (2023). *Gartner hype cycle.* https://www.gartner.com/en/chat/gartner-hype-cycle?utm_source=google&utm_medium=cpc&utm_campaign=GTR_NA_2022_GTR_CPC_SEM1_BRANDCAMPAIGNMQ&utm_adgroup=141653137858&utm_term=what%20is%20gartner%20hype%20cycle&ad=618363632099&matchtype=p&gad=1&gclid=EAIaIQobChMIyImh6OeU_wIVeQFMCh37gQ8vEAAYAiAAEgKBcfD_BwE

Greenhalgh, T., Robert, G., Macfarlane, F., & Kyriakidou, O. (2005). *Diffusion of innovations in health service organizations: A systematic literature review.* Blackwell.

Hines, A., & Bishop, P. (2015). *Thinking about the future: Guidelines for strategic foresight.* Hinesight.

Howlett, M., Ramesh, M., & Perl, A. (2009). *Studying public policy: Policy cycles and subsystems.* Oxford University Press.

Kingdon, J. W. (2011). *Agendas, alternatives, and public policies.* Longman.

Kirwin, E., Round, J., Bond, K., & McCabe, C. (2022). A conceptual framework for life-cycle health technology assessment. *Value in Health, 25*(7), 1116–1123. https://doi.org/10.1016/j.jval.2021.11.1373

Kuosa, T. (2011). *Practising strategic foresight in government: The cases of Finland, Singapore and the European Union.* S. Rajaratnam School of International Studies.

Lalonde, M. (1974). *A new perspective on the health of Canadians: A working document.* Governent of Canada. http://www.phac-aspc.gc.ca/ph-sp/pdf/perspect-eng.pdf

Lasswell, H. D. (1958). *Politics: Who gets what, when and how.* Meridian Books.

Lazar, H., Lavis, J. N., Forest, P.-G., & Church, J. (Eds.). (2013). *Paradigm freeze: Why is it so hard to reform health-care policy in Canada.* McGill-Queen's University Press.

MacNeil, M., Juzwishin, D., & Stolee, P. (2021). Navigating policy, regulatory, and health system landscapes. In A. Sixsmith, J. Sixsmith, A. Mihailidis, & M. Fang (Eds.), *Knowledge, innovation, and impact: A guide for the engaged health researcher* (pp. 305–13). Springer. https://doi.org/10.1007/978-3-030-34390-3_41

MacNeil, M., Koch, M., Kuspinar, A., Juzwishin, D., Lehoux, P., & Stolee, P. (2019). Enabling health technology innovation in Canada: Barriers and facilitators in policy and regulatory processes. *Health Policy, 123*(2), 203–214. https://doi.org/10.1016/j.healthpol.2018.09.018

McNeil, H., Elliott, J., Huson, K., Ashbourne, J., Heckman, G., Walker, J., & Stolee, P. (2016). Engaging older adults in healthcare research and planning: A realist synthesis. *Research Involvement and Engagement, 2*(1), Article no. 10. https://doi.org/10.1186/s40900-016-0022-2

Rousseau, D. M. (2020). Making evidence-based organizational decisions in an uncertain world. *Organizational Dynamics, 49*(1), 100756. https://doi.org/10.1016/j.orgdyn.2020.100756

Schneider, E. C., Shah, A., Doty, M. M., Tikkanen, R., Fields, K., & Williams II, R. D. (2021, August 4). *Mirror, mirror 2021: Reflecting poorly. Health Care in the U.S. compared to other high-income countries.* The Commonwealth Fund. https://www.commonwealthfund.org/sites/default/files/2021-08/Schneider_Mirror_Mirror_2021.pdf

Sixsmith, A., Sixsmith, J., Mihailidis, A., & Fang, M. (Eds.) (2021). *Knowledge, innovation, and impact: A guide for the engaged health researcher.* Springer. https://doi.org/10.1007/978-3-030-34390-3

United Nations Educational, Scientific, and Cultural Organization (UNESCO). (2021). *Futures literacy: An essential competency for the 21st century.* https://en.unesco.org/futuresliteracy/about#:~:text=What%20is%20Futures%20Literacy%20(FL,and%20invent%20as%20changes%20occur)

Van der Bijl-Brouwer, M. (2019). Problem framing expertise in public and social innovation. *She Ji: THe Journal of Design, Economics, and Innovation, 5*(1), 29–43. https://doi.org/10.1016/j.sheji.2019.01.003

Voros, J. (2017, February 24). The futures cone, use and history. *The Voroscope.* https://thevoroscope.com/2017/02/24/the-futures-cone-use-and-history/

White-Chu, F. E., Graves, W. J., Godfrey, S. M., Bonner, A., & Sloane, P. (2009). Beyond the medical model: The culture change revolution in long-term care. *Journal of the American Medical Directors Association, 10*(6), 370–378. https://doi.org/10.1016/j.jamda.2009.04.004

Part II

Part Two

Health Care and Health Service Delivery for the Home Environment

4

The Challenge

Older citizens have strong preferences to *age* in their homes and communities. Progressive *frailty* and *adverse health* incidents, some following time spent in *hospitals* and *rehabilitation facilities*, make it difficult to satisfy these preferences. Key challenges are to anticipate and to provide, with the help of *technological devices and systems*, effective health care in the home that is manageable by older citizens and their *health care providers*, including family members (Fig. 4.1).

Persona and Scenario—Brian and Clea

The couple, Brian and Clea, are in their eighties, having lived in their present two-story home for several decades, and enjoying their garden, community and friends. They have a married daughter with grown up children, who resides in the same city, about half an hour's drive away. The couple have modest financial means, with small private and government pensions being their primary sources of income. They live frugally, having neither a mortgage nor debts.

Clea is starting to develop dementia, being unable to remember basic facts, starting to wander, and feeling increasingly insecure. Similarly, Brian's frailty is increasing. After a severe illness resulting in the amputation of his lower left leg, his diabetes is now under control, provided he monitors his blood sugar and takes his medication as needed. However, his arthritis makes these tasks difficult. It is also hard for him to do regular house and garden work.

Consequently, neither Brian nor Clea can take care of each other properly, prepare and have nutritious meals at the right times, look after their personal hygiene and health needs,

D. Juzwishin et al., *AgeTech Innovations in Healthcare for Older Adults*, Synthesis Lectures on Technology and Health, https://doi.org/10.1007/978-3-031-86885-6_4

Fig. 4.1 Challenges of interfacing with digital devices in home environments: readability, comprehension, clutter, and distractions. *Credit* Age without limits—Age-Positive Image Library (https://www.agewithoutlimits.org/image-library)

and live safely. This is of concern to them as well as to their daughter and her family. They considered the possibilities of Brian and Clea going into a long-term care (LTC) home or the two families moving together. Both possibilities were rejected by all—at least for now.

The hope is that assistive technologies and systems can enable Brian and Clea to continue living in their home, being in contact and helped remotely, as needed, by their daughter and/or external caregivers. Medication, nutrition, personal hygiene, accident detection (especially fall detection) and wandering are everyday challenges. The COVID-19 pandemic has increased their isolation, and the use of test kits has been difficult due to their complex instructions, some provided in miniscule print (size 6 font). Safety, including fire prevention and detection, accidental burns and home security are additional concerns.

The availability of technological devices to address the needs of Brian and Clea appears very limited and fails to address their routine and special issues. Fire alarm systems and fall detection systems can be purchased but their reliability and privacy are in question. Their cost is also a major concern. It is recognized that the technological devices would need to be installed by professionals and regularly tested to ensure their operability. It is unlikely that Brian and Clea or their family members or friends could do this.

Even if suitable devices could be found and paid for, they are unlikely to meet the progressive needs of Brian and Clea without ongoing support and a system staffed by professional health care and social workers with technical capabilities.

Countries around the world demonstrate a wide variety of approaches in respect to providing health and social care to older adults to help them live in the settings of their choosing for as long as practically possible. In this chapter we focus on **h**ealth care and **h**ealth service delivery for the **h**ome environment (H3) aimed at preserving *independent* living at home. The goal is to maintain good health and manage, in particular, infirmities resulting from frailty and adverse medical events such as falls, strokes, heart attacks, infections and chronic diseases.

This chapter takes an overarching approach to home care which is expanded on in more detail in two other books in the AGE-WELL Springer series. The first is the book on *Supportive homes and communities* and the second on *Autonomy and independence.* Readers are strongly encouraged to reference those sources to get a fuller appreciation of the opportunities and challenges facing the home care sector.

Evidence confirms that citizens prefer to age with *autonomy*, *respect*, *dignity* and a *zest for life* at home, rather than in public or private institutions removed from their families, friends and neighborhoods (Home Care Ontario, 2021; Lafortune et al., 2015; March of Dimes, 2021; Philippon et al., 2021; Sharratt, 2021; VanderBent & Kuchta, 2010). Home care differs from its institutional counterparts in that the preservation of independence requires anticipating and preparing for frailty and adverse medical events *before* they occur. However, such early preparations are often lacking. Furthermore, patients are frequently discharged from acute care and rehabilitation facilities directly into their homes, which are inadequately equipped.

Home care service delivery can benefit from the development and use of appropriate technologies, especially digital technologies, provided they are developed collaboratively by technology providers and innovators in partnership with care recipients and caregivers, forming an integrated system of health preservation and health care.

While most current home care technologies are still deficient in terms of functionality, useability, availability and cost, future technologies have considerable potential. In addition to serving care recipients and caregivers more efficiently, they offer important opportunities for research and economic development, all requiring highest standards of reliability, ethics and protection of privacy. Technologies have the potential to optimize the use of scarce resources in public care systems more effectively and to create new high-quality employment and economic opportunities. These technologies are a large and growing market sector throughout the world due to increasing numbers of aging citizens and their rising expectations.

This chapter is focused on home technologies and systems that have high digital content or are digitally enabled. Home design, elevators, stairs, bathroom facilities, furnishings and similar facilities with low digital content are excluded. They are addressed in other books in Springer's AgeTech series that are identified in the Preface.

This chapter starts with important definitions and a high-level review of the literature. Together, they lead to insights and the identification of key characteristics of successful H3 technologies and systems. This is followed by ideas for some new technologies, the improvement of existing technologies and system changes. The ideas are neither exhaustive nor highly detailed. They are intended to stimulate creative thinking. The chapter concludes with high-level observations on market assessment of H3 technologies and a proposal for large seminal demonstration projects. The projects are aimed at advancing H3 technologies and their applications for the benefit of older citizens and their caregivers throughout the world.

Frailty and adverse medical events are intimately connected and manifest themselves in the impairment of one or more of the five senses (sight, hearing, touch, taste and smell), cognitive abilities and skeletomuscular functions. There is no universally accepted definition of frailty, but the following are suitable in the context of home care technologies:

> Frailty is a dynamic state affecting an individual who experiences losses in one or more domains of human functioning (physical, psychological and social), which is caused by the influence of a range of variables, and which increases the risk of adverse outcomes. (Gobbens et al., 2010)

and

> Frailty is the cumulative effect of multiple deficits in multiple domains which accumulate over time, resulting in a high risk of adverse outcomes. (St. John et al., 2019)

Similarly, the five senses, cognitive abilities and skeletomuscular functions lack universally accepted definitions. New research is deepening our understanding and it is encouraging the development of corresponding assistive devices.

AgeTech and Home Care

Our persona and an assessment of the emerging literature have led us to the importance of the following anticipatory thinking that will be expected of aging older persons, policymakers, health care providers, innovators and industry in making AgeTech in the home successful.

- Aging citizens *anticipating and preparing themselves* for frailty and health problems by becoming acquainted with what technologies are available, from where, and at what cost *before* the technologies are needed. Advance preparation also includes making financial provisions for assistive technologies since many are not covered by public health care and social services.

- In-home and outside caregivers being familiar with the range of technologies that may be needed *before* they are needed.
- *Learning* how to use technologies. Learning may be difficult, requiring visual, audio, manual and cognitive capabilities as well as special expertise, particularly computer expertise, that care recipients no longer have or have difficulties acquiring. Similarly, caregivers often have difficulties mastering specific technologies.
- *Useability*, i.e., care recipients managing the technologies under conditions of progressing frailty and over long periods of time. When self-management is limited or impossible, the technologies must be operable by in-home caregivers (like family members and close friends), placing only manageable demands on them, or on outside caregivers providing services in the home or from remote locations.
- *Reliability*, including reliability under conditions of power failures, accidental shutdowns and device freezes, i.e., cessation of functioning, as well as signaling minimum numbers of false positive readings and alerts.
- *Affordability*, comprised of initial costs and service charges for routine use and maintenance over long periods of time.
- *Confidence and trust* of the care receivers and providers in the effectiveness of technologies and their ongoing operation.
- *Harmonized systems* of care for older persons at home requiring assistive and communications technologies, including well-educated at-home and out-of-home care providers, who are appropriately resourced and supported, all functioning in accordance with sound standards and regulations.

The insights provide guidance on improving in-home care, specifically the need for:

- Making technologies more 'user friendly' for care recipients as well as for in-home and outside caregivers. Comprehensive assessment of the strengths and weaknesses of current technologies, especially for care recipients with cognitive, emotional and physical frailties, is lacking. Examples of needed technology improvements are:
  - Better communication interfaces, such as keypads and computer keyboards, mobile phones and device controllers. The interfaces, including their buttons, are generally too small to read and control, the sequences of control commands and settings are too complex, and resets resulting from inevitable system 'crashes,' i.e., ceases to function, are difficult to achieve.
  - Increased reliability of digital technologies and digitally assisted technologies, especially health monitors and alarms.
- Growing the range of digital assistive physical devices to include, for example, devices for basic physiotherapy, personal hygiene and routine wound care.
- Enhancing the ability of outside caregivers to manage, more effectively and from offsite locations, larger numbers of care recipients in their homes, by using patient specific data and information from sources that are professional and reliable.

- Improving access to outside caregivers, thereby overcoming typical recorded messages like 'make an appointment [with the outside caregiver]' with no timeslots available for days or 'go to Emergency.' Improved access to outside caregivers is an issue that needs addressing at a system level, with input from care recipients, in-home caregivers and outside providers.
- Increasing the ability of care recipients, together with their in-home and outside providers, to anticipate and prepare for their future needs *before* the needs arise and providing them with general expertise to manage their health technologies.
- Engendering higher levels of trust and privacy protection for in-home digital and digitally assisted care technologies.
- Obtaining market information on technologies and services.
- Estimating the capital and operating cost of home care technologies and services, and their underlying causes. Current costs are typically high, beyond the reach of many care recipients and their support systems.
- Providing information to innovators, companies and governmental agencies on home care technologies, their associated markets and market outlooks. Canadian and global market opportunities are growing but, while costs are currently limiting factors, they can be expected to decline as markets grow. Market forecasts are essential to stimulate innovation, investment and deployment of new technologies.
- Advancing home care system quality by applying principles of continuous and collaborative improvement, as noted by Middel et al. (2006), to the system as a whole and their constituent technological and human components.

Examples of Improved and New Technologies and Systems

The following examples of technologies and systems address important issues, particularly useability issues that inhibit adoption, raised in the preceding sections. As stated earlier, the examples are intended to stimulate thinking with the goal of benefiting care recipients and caregivers in the home environment. The examples are illustrative rather than exhaustive. Readers are encouraged to identify and consider their own situations and relevant examples.

Touch Activated Human–Machine Interfaces for Care Recipients

Most technologies require a human–machine interface, i.e., some means of communication between a person and a digital device. Frequently, the communication is by touch. For example, changing television channels, making phone calls, setting timers, adjusting thermostats and operating computers, typically requires pressing buttons, manipulating a mouse or moving a cursor. In most cases, the interfaces are too small, see photographs,

(a) (b)

Fig. 4.2 Human–machine interfaces, **a** TV channel changer and **b** portable phone. *Source* Both images of Mrs. Grace H.'s hand in Figs. 4.2(a) and (b) used with permission and gratefully acknowledged

and too touch sensitive to be managed by persons with limited dexterity and visual acuity. Similar comments apply to personal, tablet and laptop computers (Figs. 4.2 and 4.3).

These deficiencies impact usability, acceptance, safety and security. The inability to manipulate devices easily creates frustration, annoyance and often results in disuse.

In most cases, the deficiencies can be reduced by design changes that make the devices and their controls larger and less touch sensitive. However, this is rarely done and the needs of aging users are not adequately considered by technology providers. Co-design, described in Chap. 9, is one effective approach to addressing these deficiencies.

For home situations, and where space permits, preferable alternatives to existing interfaces are large touchscreens, programmed to display large buttons or keys. The touchscreens, connected to devices that require controlling or monitoring, could be wall- or table-mounted, with comfortable seating or standing areas nearby.

Large touchscreens (exceeding 1m diagonally) are already commercially available from a variety of manufacturers, including LG, NEC, Samsung, Elo and Sharp. However, they are still costly, and software is not yet readily available for home use of the type required by aging care recipients and their caregivers.

In addition to routine communications, the touchscreens can also be utilized to guide exercises and rehabilitation practices as well as provide entertainment. Good success has been achieved with touchscreen-based interventions for persons with dementia and

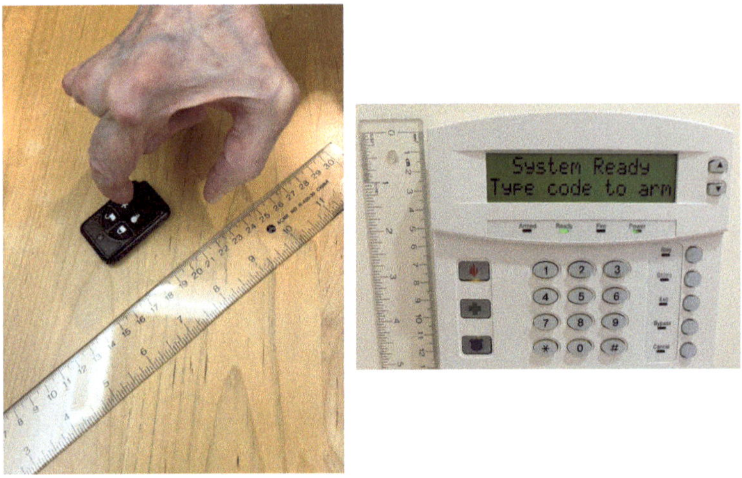

(a) (b)

Fig. 4.3 Human–machine interfaces, **a** key fob for home entry and **b** security system. *Source* The image of Mrs. Grace H.'s hand in Fig. 4.3a is used with permission and gratefully acknowledged. The image in Fig. 4.3b was taken by the author Axel Meisen

their caregivers in clinical settings. These experiences would also be applicable to home environments (Tyack & Camic, 2017).

Large touchscreens suggest major opportunities for care recipients and their caregivers as well as for technology and technology service providers.

Other Human–Machine Interfaces and Virtual Digital Assistants for Care Recipients and Caregivers

Alternatives to enlarged touch-enabled human–machine interfaces are sound- or voice-activated forms of communication. They are easy to install and to use but suffer from a lack of specificity. They can be activated by extraneous sounds, thereby creating possible confusion and safety issues. They should only be used for non-essential applications.

More versatile and effective are 'virtual digital assistants', which respond not only to sounds but are capable of interpreting spoken words since they are integrated with computing devices, programmed to process information and create response actions. Virtual digital assistants that are commercially available include names such as Siri, Alexa, and Cortona.

When the inputs are spoken words (allowing for accents and different languages), the responses can be in audio, graphical, or written forms. When the inputs are signals, such a blood pressure or electrocardiogram signals, the responses can also include alerts for

Table 4.1 Critical requirements and typical problems of virtual digital assistants

	Requirement	Typical problems
(1)	Proximity to *care recipients*	Care recipients often do not keep *virtual digital assistants* nearby
(2)	Adequate battery charge	Recharging is not done, thereby rendering the *virtual digital assistant* ineffective
(3)	Appropriate programing	Initial programing and re-programing (especially after freezes) are difficult and often beyond the capabilities of *care recipients* and *caregivers*
(4)	Reliability of *communication networks*	Interruptions and lack of *communication* are still common, especially in rural areas
(5)	Availability and expertise of the *monitoring caregivers*	*Caregivers* with good knowledge of care recipients are frequently unavailable and/or have limited expertise providing expert care, especially when conditions of *care recipients* are changing

care recipients and their caregivers, who may be on site or in remote locations. Virtual digital assistants can be programed to provide reminders for a wide variety of activities, including taking medications and engaging in exercises. In case of falls and other unusual incidents, they can send distress alerts to caregivers.

The effectiveness of virtual digital assistants depends on the requirements, shown in the Table 4.1, being met.

The requirements of Virtual Digital Assistants are difficult to meet consistently, even when network quality is adequate. Requirements (3) and (5) are particularly difficult to satisfy. Programing and reprograming typically necessitates specialized computer and systems skills that most care recipients and caregivers do not have or have difficulties acquiring. These problems can be overcome by initial and refresher training, but good training is frequently lacking and costly. It is expected that artificial intelligence (AI) will, in future, help to address these deficiencies.

Since virtual digital assistants depend on wireless and Internet communications, they are subject to targeted interference and data theft. High levels of data security and issues of data ownership are still unresolved. Major improvements are required before virtual digital assistants can be deployed with confidence and on a large scale for at-risk populations, such as older citizens with frailties.

Magnification and Other Visualization Technologies

Aging care recipients and their caregivers rely heavily on written product information. In the case of medicines and foods, printed information is often difficult to read due to the small font size. A contemporary example are the printed instructions for COVID-19 antigen test kits, marketed by Artron. However, the instructions are supplemented by a good videoclip (https://www.youtube.com/watch?v=FDzkdS3Z4sw).

Increasingly, mobile phones have magnification apps, but their practical usefulness would benefit from improvements in magnification power and display size. The devices also must be held in a fixed position for the text to be legible. This is difficult when users experience Essential Tremor(Louis, 2016; Putzko et al., 2006), Parkinson's disease, or other dexterity issues. Motion compensating technologies are available for large cameras but still require significant development for small cameras embedded in mobile phones (Kim et al., 2020).

Alternatives to magnifying small text are QR (quick response) codes that can be scanned and then displayed on mobile phones. In addition to displaying basic information stored in QR codes, links can be provided to websites that give additional information. This information can be conveyed in written or spoken form, including in several languages. The power of QR codes remains to be fully exploited in home care settings.

More advanced but also more costly communication devices are radio frequency identification (RFID) tags. In addition to providing information, they can be used to determine the location of objects to which they are attached. This is particularly helpful when care recipients wish to locate important items, such as medications, eyeglasses and hearing aids (Behera, 2022).

Mechanical Assistive Technologies

The example technologies described thus far are primarily focused on providing and exchanging information. This is valuable but many important functions required by care recipients also need mechanical interventions. Typical examples include care recipients lifting and moving objects, personal hygiene and basic wound care, preparing meals, taking medications, doing laundry, folding and ironing clothes, dressing and undressing, changing bed linens, house cleaning, making minor repairs and hobbies such as gardening. All these activities require a complex, interactive combination of sensing, data processing and mechanical actuation. The activities must be performed under different conditions, without exact repetition. These widely different and varying characteristics distinguish home-based activities from those performed, for example, in factory-based manufacturing which have a high degree of repetition and where automation, using basic robotic principles, is more readily applicable.

Good summaries and examples of mechanical assistive devices with important digital content applicable to home care are provided by Ishigami et al. (2021); Mohammed et al. (2015); Morsi et al. (2016); Moustris et al. (2021); Reparaz et al. (2019); and Seo and Lee (2019). Digitally enhanced mechanical assistive devices cannot, at present, fulfill most tasks required by home-based care recipients. The diversity of tasks to be performed, lack of structure in the home environment, and complex human physiology are primary challenges, but the lack of appropriate sensors and actuators are also major technological barriers. It is appropriate to focus efforts on the development of devices that address important, manageable problems with wide applicability. Assistive devices that meet personal hygiene needs are particularly needed and promising. They also have large market potential, an important precondition for development and investment by private and governmental organizations (Blackman, 2013).

System Technologies for Remote Care Monitoring and Provision

Technologies that enable remote monitoring of care recipients with frailties at home and providing them with remote support already exist to some extent. They are becoming more sophisticated than relaying simple alerts resulting from emergencies, like falls. The recent pandemic has stimulated and seems to prove the efficacy of telehealth when properly instituted and coordinated. Babylon by TELUS Health (TELUS Health MyCare™), CloudMD, Teladoc, Maple and the Ontario Telemedicine Network are examples of telehealth providers in Canada. Silverbullet is a good example used in Denmark. However, more evidence needs to be provided and more rigorous assessments are necessary. Earlier initiatives undertaken in the United Kingdom indicate caution about drawing conclusions prematurely (Hendy et al., 2012, 2013).

The expertise and capabilities of telehealth systems are also applicable to aspects of home care that are focused on well-being and managing daily activities. Key issues are usability, reliability and affordability. Usability remains a primary issue since the interfaces and instructions are typically too complex for many care recipients and their care providers. More user-friendly training and retraining programs are needed.

An important issue is data security, which was already mentioned in connection with virtual digital assistants. The problem is even more important for system technologies since security breaches potentially affect all care recipients and care providers associated with a particular system. Hacking the data and communication networks of entire organizations and all of their subscribers, rather than hacking individual devices, is a primary concern. Security can only be addressed at organizational and network levels. This is a major development and market opportunity for digital systems companies. For reasons of history, geography, culture, commitment to universal health care and innovation, Canada is well-positioned for the growth of companies and organizations in the H3 market sector. Closely related to data security is data ownership. It is presently unclear and legislation or regulations may be needed to establish data ownership.

Anticipatory Digital Technology Systems

The digital technology systems for home care described in the literature and the preceding sections are primarily focused on *maintaining* a satisfactory quality-of-life under conditions of frailty. The systems are not focused on addressing and making provisions for managing *future* frailties that may be more diverse and severe. Systems are needed to anticipate these needs and prepare care recipients and care providers for managing them.

While the future, including the future conditions of care recipients, cannot be predicted with certainty, it is possible to anticipate them based on the characteristics of *individual* care receivers and their circumstances. Artificial intelligence provides a means of anticipating future conditions and thereby preparing for them. Such anticipatory systems could build on the expertise developed by large companies engaged in medical data analytics, such as IBM Watson Health and Microsoft Cloud for Healthcare. However, the scope for medium and small companies to provide valuable services in the anticipatory H3 sector is also large. What is needed are innovative approaches that are respectful of ethical, legal and personal concerns and that are based on rigorous sciences.

Comprehensive Systems

While the preceding sections focused on technological and communication devices, it is important to emphasize that these devices are part of comprehensive systems that include human components (particularly care recipients and caregivers) as well as organizational, funding and educational components. Unless these components are aligned and fully functional, the desired outcomes for care recipients cannot be achieved.

In many countries, comprehensive systems are lacking. The reasons are numerous and include divided jurisdictions, shortages of system-wide funding, low or no-pay for care providers, inadequate regulations and standards, and the absence of educational and training programs. These challenges cannot be resolved by technological developments alone but require public- and private-sector leaders with expertise in current and emerging H3 issues.

The private sector's role in the home care sector is still small, largely because its economic growth potential has not yet been fully recognized and assistive device development and provision are largely the purview of small- and medium-size companies. However, age demographics and the preferences of older citizens to age in their homes suggest that the market will grow rapidly and become large.

Seminal Demonstration Project

As shown in the previous sections of this chapter, numerous initiatives in the H3 sector are already underway and new initiatives will undoubtedly be conceptualized. These developments are occurring in many countries throughout the world but, while cultural and local differences are important and must be addressed, unnecessary duplication and gaps are also visible. To make better use of limited funding and expertise, and to accelerate progress, a coordinated approach is proposed.

The proposed approach is an annual *Innovation forum on H3 technologies and systems*, an initiative aimed at fostering, demonstrating and validating seminal ideas for improved and new equipment and systems.

Since age-related frailties are a global challenge, the *Innovation forum* should be organized in collaboration with the World Health Organization. The United Nations General Assembly designated WHO to lead activities of the Decade of Healthy Ageing (2021–2030). The proposed *Innovation forum* is well aligned with this designation.

The principal objectives of the *Innovation forum* would be:

- To identify the needs for major advances in existing and fundamentally new home care technologies and systems that are technically feasible, effective, robust and user friendly, affordable and applicable to a wide range of home settings.

The learnings from previous international meetings, such as the GREAT summit (Smith et al., 2018) should be taken into account. The Design for Dignity (D4D) Model advocated for by Langley et al. (2020) provides a logical construct for identifying and developing technologies that support the quality-of-life of older citizens in their home environments.

- To assign different countries or groups of countries the lead on developing specific technologies and systems, thereby creating a critical mass of development effort, while avoiding unnecessary duplication. Special attention should be given to the needs and experiences of Low Income countries, such as those recently reported for Tajikistan (Mishra et al., 2021).
- To involve care recipients, in-home caregivers and outside caregivers as well as representatives of technology development organizations, manufacturers, health and care organizations and funders in the *Innovation Forum* to ensure relevance and steady progression to development, trials and full-scale implementation.
- To report on the home care technologies and systems identified and conceptualized, with the view to supporting continuous progress and wider uptake by care recipients and caregivers.

The *Innovation forums* should be conducted annually in different regions of the world, with each regional *Innovation forum* focused on a specific group of frailties and corresponding assistive technologies and systems. This approach would maximize outcomes and minimize duplication of efforts, provided there is strong coordination between individual efforts and conclusions are based on comprehensive data and rigorous analyses.

While all health technologies require harmonization with local conditions, norms and regulations, there are sufficient common elements to make global approaches feasible.

Concluding Remarks

We conclude this chapter with the persona described at the start of this chapter, i.e., the circumstances of Brian, Clea, their family and caregivers. They can now be revisited in light of available and needed technologies. The following is clear:

- Technologies for self- and remote-monitoring are available but, except for basic alarm technologies, are typically too complex and likely too costly to be acquired and maintained. The same applies to communication technologies.
- In addition to the lack of accessible monitoring technologies, concerns about privacy, 'being watched all the time', and data ownership are of concern and remain largely unresolved.
- Home care technologies are unavailable to support many daily needs regarding nutrition, hygiene, medication and other personal tasks, including management of homes and gardens. These needs can be expected to grow as dementia and physical disabilities increase.
- There is a dearth of satisfactory digital technologies that enable Brian, Clea and their daughter to plan and prepare for a future that will likely present even greater challenges of frailty and debility.
- A comprehensive system consisting of appropriate technologies, experienced care providers, and funding sources is lacking and needs to be developed to meet the current and emerging needs of older citizens like Brian and Clea. The system must also enable their children and other family members to provide care at home.

The lack of suitable and accessible home care technologies should not only be seen as a problem but as an encouragement to develop such technologies and make them widely available in Canada and throughout the world. This would immensely improve the quality-of-life of older adults while also providing greater satisfaction for caregivers and new opportunities for researchers, innovators and businesses.

References

Behera, S. K. (2022). Chipless RFID sensors for wearable applications: A review. *IEEE Sensors Journal, 22*(2), 1105–1120. https://doi.org/10.1109/JSEN.2021.3126487

Blackman, T. (2013). Care robots for the supermarket shelf: A product gap in assistive technologies. *Ageing & Society, 33*(5), 763–781. https://doi.org/10.1017/S0144686X1200027X

Gobbens, R. J., Luijkx, K. G., Wijnen-Sponselee, M. T., & Schols, J. M. (2010). Toward a conceptual definition of frail community dwelling older people. *Nursing Outlook, 58*, 76–86. https://doi.org/10.1016/j.outlook.2009.09.005

Hendy, J., Chrysanthaki, T., Barlow, J., Knapp, M., Rogers, A., Sanders, C., Bower, P., Bowen, R., Fitzpatrick, R., Bardsley, M., & Newman, S. (2012). An organisational analysis of the implementation of telecare and telehealth: The whole systems demonstrator. *BCM Health Services Research, 12*, 403–413. https://doi.org/10.1186/1472-6963-12-403

Hendy, J., Barlow, J., & Chrysanthaki, T. (2013). Stimulating whole system redesign: Lessons from an organizational analysis of the Whole System Demonstrator programme. *Journal of Health Services Research & Policy, 18*(Suppl I), 47–55.

Home Care Ontario (2021, October 8). 97 per cent of Ontario seniors want increased home care funding: Poll, Home Care Ontario. *Newswire.ca*. https://www.newswire.ca/news-releases/97-per-cent-of-ontario-seniors-want-increased-home-care-funding-poll-805034645.html

Ishigami, Y., Jutai, J., & Kirkland, S. (2021). Assistive device use among community-dwelling older adults: A profile of Canadians using hearing, vision, and mobility devices in the Canadian longitudinal study on aging. *Canadian Journal on Aging/la Revue Canadienne Du Vieillissement, 40*(1), 23–38. https://doi.org/10.1017/S0714980819000692

Kim, W.-Y., Seo, H.-T., Kim, S., & Kim, K.-S. (2020). Practical approach for controlling optical image stabilization system. *International Journal of Control, Automation and Systems, 18*(4), 824–833. https://doi.org/10.1007/s12555-018-0913-0

Lafortune, C., Huson, K., Santi, S., & Stolee, P. (2015). Community-based primary health care for older adults: A qualitative study of the perceptions of clients, caregivers and health caregivers. *BMC Geriatrics, 15*, 57. https://doi.org/10.1186/s12877-015-0052-x

Langley, J., Wheeler, G., Partridge, R., Bec, R., Wolstenholme, D., & Sproson, L. (2020). Designing with and for older people. In A. Woodcock, L. Moody, D. McDonagh, A. Jain, & L. C. Jain (Eds.), *Design of assistive technology for ageing populations* (pp. 3–19). Intelligent systems reference library, Vol. 167. Springer Nature Switzerland AG.

Louis, E. D. (2016). Essential tremor: A common disorder of Purkinje neurons? *The Neuroscientist, 22*(2), 108–118. https://doi.org/10.1177/1073858415590351

March of Dimes Canada. (2021, April 27). National survey shows Canadians overwhelmingly want to age at home; Just one-quarter of seniors expect to do so. *Newswire.ca*. https://www.newswire.ca/news-releases/national-survey-shows-canadians-overwhelmingly-want-to-age-at-home-just-one-quarter-of-seniors-expect-to-do-so-842023628.html

Middel, H., Boer, H., & Fisscher, O. (2006). Continuous improvement and collaborative improvement: Similarities and differences. *Creativity and Innovation Management, 15*(4), 338–347. https://doi.org/10.1111/j.1467-8691.2006.00407

Mishra, S., Pupulin, A., Ekman, B., Khasnabis, C., Allen, M., & Huber, M. (2021). National priority assistive product list development in low resource countries: Lessons learned from Tajikistan. *Disability and Rehabilitation: Assistive Technology, 16*(8), 857–864. https://doi.org/10.1080/17483107.2020.1745908

Mohammed, S., Moreno, J. C., Kong, K., & Amirat, Y. (Eds.). (2015). *Intelligent assistive robots: Recent advances in assistive robotics for everyday activities*. Springer tracts in advanced robotics,

Vol. 106. Springer International Publishing Switzerland. https://doi.org/10.1007/978-3-319-12922-8

Morsi, Y. S., Shukla, A., & Rathore, C. P. (2016). *Optimizing assistive technologies for aging populations.* IGI Global. https://doi.org/10.4018/978-1-4666-9530-6

Moustris, G., Kardaris, N., Tsiami, A., Chalvatzaki, G., Koutras, P., Dometios, A., Oikonomou, P., Tzafestas, C., Maragos, P., Efthimiou, E., Papageorgiou, X., Fotinea, S.-E., Koumpouros, Y., Vacalopoulou, A., Karavasili, A., Nikolakakis, A., Karaiskos, K., & Mavridis, P. (2021). Human-friendly robotics 2020, Springer proceedings in advanced robotics. In B. Siciliano & O. Khatib (Eds.) *The I-Walk assistive robot—A multimodal intelligent robotic rollator providing cognitive and mobility assistance to the elderly and motor-impaired.*

Philippon, D., McPherson, A., Fyke, K., Johnson, M., Juzwishin, D., McKendrick, W., Mark, K., Meisel, A., Miller, L., & Roger, R. (2021, May 22). We need a national shift to ensure Canadians age with 'zest and dignity'. *Calgary Herald.* https://calgaryherald.com/opinion/columnists/opinion-we-need-a-national-shift-to-ensure-canadians-age-with-zest-and-dignity

Putzke, J. D., Whaley, N. R., Baba, Y., Wszolek, Z. K., & Uitti, R. J. (2006). Essential tremor: Predictors of disease progression in a clinical cohort. *Journal of Neurology, Neurosurgery & Psychiatry, 77*(11), 1235–1237. https://doi.org/10.1136/jnnp.2006.086579

Reparaz, M. J. T., Gorroñogoitia, A., & Odriozola, J. (2019). Comparative analysis of assistive technologies for older people: Hoist or patients lifts. *International Journal of Integrated Care, 19*(S1), A328, 1–8. https://doi.org/10.5334/ijic.s3328

Seo, H., & Lee, S. (2019). Design of general-purpose assistive exoskeleton robot controller for upper limbs. *Journal of Mechanical Science and Technology, 33*, 3509–3519. https://doi.org/10.1007/s12206-019-0645-y

Sharratt, A., (2021, November 16). Amid soaring prices, retired Canadians are staying in their homes: 'This is a wake-up call for all of us'. *Globe and Mail.* https://www.theglobeandmail.com/investing/personal-finance/retirement/article-amid-soaring-prices-retired-canadians-are-staying-in-their-homes-this/#:~:text=%E2%80%9CMost%20of%20us%20want%20to,2021%20poll%20by%20Campaign%20Research.

Smith, R. O., Scherer, M. J., Cooper, R., Bell, D., David A., Hobbs, D. A., Pettersson, C., Seymour, N., Borg, J., Johnson, M. J., Joseph P., Lane, J. P., Sujatha, S., Rao, P. V. M., Obiedat, Q. M., MacLachlan, M., & Bauer, S. (2018). Assistive technology products: A position paper from the first global research, innovation, and education on assistive technology (GREAT) summit. *Disability and Rehabilitation: Assistive Technology, 13*(5), 473–485. https://doi.org/10.1080/17483107.2018.1473895

St. John, P. D., McClement, S. S., Swift, A. U., & Tate, R. B. (2019). Older men's definitions of frailty—The Manitoba follow-up study. *Canadian Journal of Aging, 38*(1), 13–20. https://doi.org/10.1017/S0714980818000405

Tyack, C., & Camic, P. M. (2017). Touchscreen interventions and the well-being of people with dementia and caregivers: A systematic review. *International Psychogeriatrics, 29*(8), 1261–1280. https://doi.org/10.1017/S1041610217000667

VanderBent, S., Kuchta, B., & Community Providers Association Committee. (2010, November). *Valuing home and community care.* Community Providers Association Committee. https://www.homecareontario.ca/docs/default-source/publications-mo/valuing-home-and-community-care.pdf?sfvrsn=4

Long Term Residential Care 5

The Challenge

Most older persons wish to remain in their own homes for as long as possible. Unfortunately, health and social care systems in many countries have not evolved to provide adequate or sufficient community-based alternatives to long term residential care. Continued lack of such alternatives reflects policy and system priorities that could be considered ageist, resulting in many older persons being admitted to a care setting they would not prefer. Long term care homes provide care for aging and increasingly at risk and diverse resident populations, but face significant challenges related to funding, staffing and care quality. These deficiencies were highlighted and exacerbated by the COVID-19 pandemic, with high rates of COVID-19 mortality in many countries. Social isolation and loneliness are common issues for older adults in LTC. The well-being and quality-of-life of many LTC residents are compromised by outdated and overcrowded facilities. More positive outcomes can be achieved with robust policy responses and increased resources, and innovative care models and new technologies give cause for optimism. The focus of this chapter is on the potential for AgeTech to support the care and quality-of-life of older residents of LTC homes (Fig. 5.1).

Persona and Scenario—Agnes

Agnes is an 89-year-old widow. For the past six months, she has been living in a 95 bed LTC home in a mid-sized city. The home is owned and operated by a small for-profit chain of retirement communities. Agnes has a daughter, Lauren. Shortly after Agnes moved into the LTC home, Lauren and her family moved to a city three hours away to be closer to Lauren's husband's work. The family has been unable to find a LTC bed in their area

D. Juzwishin et al., *AgeTech Innovations in Healthcare for Older Adults*, Synthesis Lectures on Technology and Health, https://doi.org/10.1007/978-3-031-86885-6_5

Fig. 5.1 Long-term care resident. *Credit* Age Without Limits—Age-Positive Image Library. https://www.agewithoutlimits.org/image-library

that would allow Agnes to live closer to them. Agnes is often bored and has been quite lonely since moving into the home. Lauren and her family are unable to visit often but try to stay in touch with telephone calls. Agnes has severe arthritis and associated pain; she has been using a walker since she fractured her hip after a fall in her doctor's office. She was diagnosed with Alzheimer's disease a year ago. Agnes needs assistance with daily activities, including getting out of bed, toileting, dressing and bathing. She has had frequent episodes of incontinence and staff are often too busy to take her to the bathroom or to change her incontinence undergarments. Agnes is beginning to develop pressure ulcers. She receives 12 prescription medications to manage her arthritis and other chronic conditions. She has been prescribed PRN (as needed) medication to alleviate her arthritis pain, but Agnes often does not verbalize her pain to staff, feeling she does not want "to be a bother."

Long Term Residential Care

Long Term Residential Care
Although most older adults would prefer to remain in their own homes, many require long term residential care due to a lack of community-based alternatives.

LTC homes provide care for aging and increasingly at risk and diverse resident populations, but face significant challenges related to funding, staffing and care quality. In many jurisdictions, these challenges were exacerbated during the COVID-19 pandemic.

LTC systems vary internationally in terms of organization, financing, staffing, and number of beds per population aged 65+. The physical environment of LTC homes plays a role in resident well-being, but many homes are aging and over-crowded. Improving long-term residential care is an ethical and financial imperative.

Older adults who are no longer able to live independently in their own homes may require admission to LTC homes, commonly called "nursing homes", where they can receive 24-h supervision, support, and nursing care (Ontario Ministry of Long-Term Care, 2024).

We define LTC homes as long term residential settings to which older persons may be admitted for ongoing care. For the purposes of this chapter, we distinguish long term residential care from care that may be provided on a long term basis in the older person's own home—a focus of Chap. 2. We also distinguish long term residential care settings from retirement homes. Although retirement homes are residential settings in which older persons may live for a longer term, we would consider the retirement home to be the residents' own home in the community. Additionally, retirement home residents may not require ongoing care. We note, however, that recent work has shown that many retirement home residents receive levels of care and services comparable to that received by LTC residents, thus some issues and opportunities in that sector may be similar (Roblin et al., 2019).

International Variation
LTC systems vary internationally in their organization and financing (Dyer et al., 2019). Some countries rely heavily on family caregivers or mandate that families must provide LTC for older people (Dyer et al., 2019). Funding sources vary. As examples, in Australia and Denmark, nearly all (90%+) of LTC is publicly funded. In Germany and the Netherlands, LTC is predominantly financed through social insurance. In Canada most LTC funding comes from a combination of public funding and resident fees (public health insurance (~ 70%) and resident co-payment). Japan has a roughly equal split

between social insurance and general taxes. In Poland, nearly all LTC services are privately financed (Dyer et al., 2019). In most jurisdictions, a portion of LTC costs is covered by out-of-pocket expenditure (Dyer et al., 2019).

Institutional care in some countries has roots in French and English welfare systems, including the *English Poor Law of 1601* and its associated workhouses (Emodi, 1977; Forbes et al., 1987). It has been observed that this development "was affected by the prevailing attitude in the eighteenth and nineteenth centuries that those in need of assistance were largely to blame for their situation" (Forbes et al., 1987, p. 15). That "institutional eldercare has its origins in the penal system," as noted by Picard (2021, p. 30), was a foreshadowing of its current low status and priority in the health care system.

Internationally, there is considerable variation in the number of LTC beds per population of people aged 65 and over (OECD, 2017) and in the proportion of the older population housed in institutions (Dyer et al., 2019). The Organisation for Economic Co-operation and Development (OECD) reported an average of 50 LTC beds per 1,000 people 65+in 2015, ranging from 87/1000 in the Netherlands to fewer than 20/1000 in Italy, Latvia, Poland and Turkey (OECD, 2017). While these ratios have remained fairly stable over time, some countries have been able to reduce LTC beds as a result of more comprehensive care systems and strengthened community resources (OECD, 2017).

New Models

A variety of new models for LTC have been proposed (Welsh, 2021), however the evidence for these models is inconsistent. The Green House Model (Power & Carson, 2021), has a "small house" home-like model with no more than 12 residents, and with the older residents having greater control over their daily lives, and self-managed teams of empowered staff (Zimmerman et al., 2021). This model has been evaluated with some evidence of improved outcomes (Zimmerman et al., 2016). A recent study found that homes operating with this model had fewer COVID-19 cases and deaths (Zimmerman et al., 2021). The Butterfly Household Model of Care was proposed by David Sheard of the United Kingdom as a person-centered emotion-focused model of dementia care (Spurgaitis, 2018). Despite some promising anecdotal accounts of the Butterfly Model (Vogel, 2018; Welsh, 2021) and descriptions of its implementation (Versaevel, 2018), there appears to be little research evidence for the effectiveness of this model (CBC News, 2018). This is a concern as principles of person- or patient-centered care do not always translate into practice (Haslam, 2013; Hodgkin & Taylor, 2013). It also seems unlikely that a philosophy of emotions-focused or person-centered care would be sufficient to overcome severe staff shortages.

LTC Homes Face Significant Challenges

LTC homes house the most at risk older adult populations but also face significant challenges related to limited funding, inadequate staffing and a poor quality of care (Jain et al., 2019).

The COVID-19 pandemic had a devastating impact on LTC, with high rates of COVID-19 mortality in LTC homes in many countries (Declercq et al., 2020; Dykgraaf et al., 2021; Estabrooks et al., 2020). In Canada, a higher proportion of COVID-19 deaths occurred in LTC homes than in other OECD countries (CIHI, 2020a). The disproportionate impact of the pandemic on LTC provides an unfortunate measure of how LTC is valued, or rather, not valued, in health care systems, as policymakers and health administrators prioritized resource allocation to other sectors such as acute care (Declercq et al., 2020). Countries such as Denmark and Germany, which responded more aggressively to the pandemic with improved staffing and infection control practices, experienced lower rates of COVID-19 mortality (Sepulveda et al., 2020). These experiences illustrate that outcomes for at risk LTC residents can be improved with adequate policy responses and resources.

While the pandemic exposed serious deficiencies in the LTC sector, these deficiencies are longstanding (Hsu et al., 2020). A review by Wong and colleagues (2021) found that between 1998 and 2020, eighty reports of LTC commissions or inquiries were produced in Canada. Their review demonstrated that the main issues in LTC have been known and well studied for many years, and that reports have yielded very consistent recommendations. According to Wong et al. (2021, p. 6), the following 12 areas were recurrent themes of the reports' recommendations:

- funding (increase or redistribute)
- LTC quality of care
- education and training
- staffing mix
- data driven (collect and use data to improve LTC)
- staff support (resources or funding for LTC staff support, including emotional support)
- behaviors (support and resources for residents with responsive behaviors)
- federal involvement
- for-profit homes (regulate, reduce or reform)
- critical incidents (reporting and tracking)
- commission (appoint commission or task force, etc.)
- infection control (establish national standard).

Many recommendations related to the need to increase funding and improve staffing.

Increasingly Complex and Diverse Resident Population

Funding and staffing of LTC homes have not kept pace with an aging, increasingly complex resident population. As an example, this profile of LTC residents is based on data from the Canadian Institute for Health Information (2020b) on the characteristics of LTC residents in Canada:

- The average age of residents is 83 years (54.3% are 85+).
- Percentage of residents who are female: 65.2%.
- Percentage of residents who require extensive assistance or are dependent on others to perform activities of daily living (ADL): 82.8%, however only 35.6% receive any physical or occupational therapy.
- Incontinence is common: 67.6% have frequent or total bladder incontinence; 51.6% have frequent or total bowel incontinence.
- Percentage of residents who have daily pain: 8.5%.
- Many are at risk for pressure ulcers: ~ 15% have a Pressure Ulcer Risk Scale score of 4 or greater, resulting in a ~ 10% or higher incidence of pressure ulcers (Poss et al., 2010).
- Over half (52.5%) have depressive symptoms.
- Most residents (81.2%) have at least mild cognitive impairment; 32.7% have severe cognitive impairment.

Social isolation and loneliness are common in older adults and are risk factors for LTC admission. LTC residents are also at increased risk for their occurrence (Freedman & Nicolle, 2020). Social isolation and loneliness for LTC residents were greatly exacerbated by the pandemic as a result of policies that required social distancing and that restricted visits from family members and friends. Social isolation is more likely to be experienced by older immigrants (Employment and Social Development Canada, 2018) or LTC residents who are geographically isolated from family and friends, such as residents living in rural homes in geographically vast countries such as Canada or Australia (Parmenter et al., 2012).

The population of LTC residents is becoming increasingly diverse in terms of race and ethnicity (Koehn et al., 2018; Martinsson et al., 2013); gender identity and sexual orientation (Fasullo et al., 2021; Stinchcombe et al., 2017; Wilson et al., 2018); and with increasing numbers of Indigenous older persons (Sue Cragg Consulting & the CLRI Team, 2017). This increasing diversity highlights the critical importance of culturally responsive approaches (Mullings & Given, 2013; Giwa et al., 2020; Parker & Geron, 2007) as well as efforts to promote equity (Fasullo et al., 2021) and to address systemic racism (Sloane et al., 2021).

While most LTC residents could benefit from palliative care, the Canadian Institute of Health Information (CIHI) reports that only 6% of LTC residents in Canada, identified as those having less than six months to live, received palliative care (CIHI, 2018).

Staffing and Workforce Issues

As noted, limited funding and inadequate staffing have been perennial issues in LTC, leading to concerns of care quality (Clemens et al., 2021); resident safety (Boscart et al., 2018; Wagner & Rust, 2008); elder abuse (Hawes, 2003); and staff burnout (Woodhead et al., 2016). Although more research is needed on the association between staffing and the quality and outcomes of care, available evidence indicates that higher staffing levels and skill mix are associated with better outcomes (Clemens et al., 2021; Twigg et al., 2021). While LTC staff: resident ratios vary considerably across countries (OECD, 2022), the recruitment and retention of skilled care staff is recognized as a global challenge (Stone, 2019). Many LTC homes find it difficult to attract and retain staff due to issues such as challenging working conditions, high workloads and a negative public image of the sector as well as lower wages than in acute care. In addition, some staff also report that high levels of documentation required for funding take away from resident care (Ministry of Long-Term Care, 2020). Lack of care capacity, including limited primary medical care, often means that residents needing medical care are transferred to emergency departments rather than given care in the home (Marshall et al., 2016).

The makeup and training of the LTC workforce varies across countries (Jain et al., 2019). In Canada, 80% of direct care to LTC residents is provided by personal support workers (care aides) with limited training (Berta et al., 2013). Most are older women who speak English as a second language; many need to work at more than one job to make a living (Chamberlain et al., 2019). The "gendered and racialized devaluing of migrant labour" (Gahwi & Walton-Roberts, 2020, p. 2) has been associated with a general devaluing of LTC (Brown & Braun, 2008). Gahwi and Walton-Roberts (2020) argue that this contributed to worsening the impact of COVID-19 in LTC homes.

Physical Environment of LTC Homes

The physical environment of the LTC home plays a major role in residents' health and well being (Armstrong & Braedly, 2016; Chaudhury et al., 2018; Marquardt et al., 2014; Power & Carson, 2021). Aging and over-crowded facilities with many shared rooms and inadequate infection control were a significant factor in the spread of COVID-19 in LTC (Betini et al., 2021; Brown et al., 2021; Declercq et al., 2020; Heckman et al., 2021). There is support for more home like residential environments (Armstrong & Braedly, 2016; Heckman et al., 2021; Levitt, 2021). There are also concerns that home-like modifications may not always support the comfort and function of residents, as in an example from Corazzini et al. (2015) where a chair selected for its home-like appearance was too low for the resident to get out of without help. Armstrong and Braedly (2016) note the potential conflict between creating a home-like atmosphere and supporting infection control measures. It should be noted also that a "home-like" physical environment does not ensure that residents are "feeling at home." Cooney (2011) found that while feeling at home could be supported or hindered by the physical environment, its core determinants related to promoting continuity in daily activities, preserving personal identity and fostering a sense of belonging.

The Ethical and Financial Imperative

Most older persons wish to continue living in their own homes for as long as possible—to "age in place" (National Institute on Aging, 2021). Moving to a LTC home is often considered a last resort (Campbell-Enns et al., 2020), often precipitated by a crisis with little advance planning or discussion (Stolee et al., 2014). Given the long-standing challenges faced by the LTC system, and with the particularly tragic experience of the COVID-19 pandemic, improvement of long term residential care is an ethical imperative.

As noted by Wong and colleagues (2021) however, policymakers have been more inclined to study these problems than to address them. The cost of improvements to the existing system are likely to be considered prohibitive by policymakers. MacDonald and colleagues (2019) project that the cost of LTC in Canada, following *current* delivery trends, would increase from $22 billion to $71 billion (in constant 2019 Canadian dollars). They further estimate that paying for currently unpaid hours of care in the home (in the community), from the public purse, would add an additional $27 billion by 2050. Implementing recommendations for system improvement under consideration by the Canadian parliament, including increasing the number of beds by 52,000, increasing wages and hours of care as well as increasing the number of hours of publicly funded home care, was estimated to cost $13.7 billion per year (Office of the Parliamentary Budget Officer, 2021). Both of these sets of cost estimates assume that the structure of the current long-term care system would remain relatively constant. In the American context, Grabowski (2021) suggests that LTC requires massive funding increases at a level similar to the war on cancer.

Kuluski and colleagues (2012) note that wait lists for LTC continue to grow across high income countries despite efforts toward deinstitutionalization. Their analysis of LTC wait lists indicates that these may inflate the apparent demand for LTC (Kuluski et al., 2012). Consistent with other research (Williams et al., 2009), they found that demand for LTC beds could be significantly reduced through strengthened home and community care options.

An ethical response to the challenges of LTC homes requires significant additional investment, but also a fundamental re-thinking of their role and structure as well as substantial investment in community care options and prevention programs.

AgeTech in Long Term Care Homes

AgeTech in Long Term Care Homes

Many technologies could have a positive impact on long term care residents and staff. Examples include:

- healthcare information and management technologies
- technologies to support communication and social interaction
- technologies for entertainment and recreation
- social robots and automated prompting systems
- remote monitoring technology, including sensors and wearables.

There has been limited adoption of technologies in long term care. Barriers include the limitations of currently available technologies, as well as system-level factors that hinder adoption.

Uptake of beneficial technologies should be supported across systems to ensure they do not exacerbate inequities for residents and staff.

Technology has a wide range of applications in LTC. Chapman and colleagues (2019), based on an environmental scan of peer-reviewed publications, grey literature and corporate websites, developed a taxonomy of categories of technologies with potential for impact on older persons or staff (this taxonomy included technologies both for clients' homes and LTC homes). Chapman and colleagues' (2019, p. 12) taxonomy includes the following categories:

- robots
- electronic health record interoperability
- education
- remote monitoring
- wearables
- family caregiver support
- online care management
- home-to-clinic communication
- staffing and recruitment
- facility/home health management
- documentation
- patient-employer engagement
- alarms and alert systems
- assisted living.

This taxonomy was used by the McMaster Health Forum (Gauvin et al., 2021) in their examination of the potential for use of technology in LTC settings in Canada. Their examination focused on technologies that support (1) communication; and (2) the provision of care and supports.

Notable by their absence in the above taxonomy are technologies that are focused on entertainment and recreation that might enhance residents' enjoyment and quality-of-life. Several of these are presented later in this chapter.

Limited Adoption of Technologies in LTC

Despite the many areas of potential application of technologies in LTC, actual application of technologies in this context has been limited (Wilson et al., 2022). In addition to technological barriers, i.e., limitations in the capacities of current technologies to undertake particular tasks, there are a variety of barriers which have hindered the development and use of technologies in LTC. The McMaster Health Forum (Gauvin et al., 2021, pp. 16–7) identified system-level factors that hinder adoption:

- *Governance arrangements*: These include the complexity of dealing with legislation and regulations from multiple jurisdictions, and a slow and challenging regulatory environment. The policy process and LTC homes are risk-averse, and complex policies such as privacy legislation can hinder use of technology.
- *Financial arrangements*: These include a variety of private and public funding arrangements, no coordination of group purchasing and limited resources in a sector reluctant to invest in technology at the possible expense of basic care.
- *Delivery arrangements*: These include limited staffing resources and infrastructure, combined with the additional demands presented by training on the new technology and incorporating it into practice and workflows. A key challenge is the gap between users and vendors. There has been limited use of partnership approaches involving residents, staff and management—homes are likely to become involved late in the technology development process, making it less likely that the technology will meet their needs or preferences, or that it can be integrated successfully with other technologies and systems in the homes.

To ensure that inequities for LTC residents and staff are not exacerbated by the introduction of technologies, uptake needs to be supported across systems (Wilson et al., 2022).

Using Technology to Support Person-Centered Care

Based on their understanding of the needs and capacities of LTC staff, residents, family care partners and of emerging evidence, Chu and colleagues (2021a) made five recommendations related to the use of technology to support person-centered care in LTC, both during and beyond the COVID-19 pandemic. These include:

- *Stable and reliable Wi-Fi access*: wireless Internet access is limited in LTC homes, which has impeded adoption of technologies such as those that could support social connectedness and reduce social isolation. Lack of wireless connectivity was a challenge in a recent study of the use of a mobile health application to support Portuguese nursing home staff (Esteves et al., 2019).
- *Software and peripheral accessories*, that support residents' independent use of technology: technologies are often not designed to be accessible for persons with physical, sensory, or cognitive impairments—usability could be enhanced by co-design and engagement processes.
- *Technology to support clinical decision-making*: use of resident data from electronic records or other sources can be used to support evidence-informed decision-making for resident care.
- *Social technologies*: technologies that support social connection and physical activity, including music technologies, video exercise or gaming systems, and virtual reality technologies, are particularly valuable during periods of social isolation (such as a pandemic).
- *Environmental control technologies*: intelligent technologies can help residents and staff to manage the LTC home environment (e.g., temperature control, turning off lights), and support resident independence and reduce demands on staff. However, they may be challenging for some residents with sensory or cognitive impairments.

Technologies that Could Help Agnes and Other LTC Residents

As outlined earlier in this chapter, there are many potential technologies that could be used in LTC. Not all of these can be explored in this chapter, however we discuss several priority areas where technologies could benefit Agnes and other LTC residents, while potentially reducing demands on staff.

Health Care Information and Management Technologies

Several of the areas identified for improvement in LTC that were identified in the review by Wong and colleagues (2021) related directly or indirectly to the use of electronic health records (EHRs) or other health information systems for administration, care planning and quality improvement. Despite their potential for benefit in LTC, adoption of health information technologies has lagged in this sector (Cherry et al., 2011; Wang et al., 2014). EHRs are discussed in greater detail in Chap. 8.

In LTC, health information technology has been used for resident care, clinical decision support, and administrative functions such as billing (Alexander et al., 2017). Historically, technology was used primarily for administrative functions, and less for resident care and clinical support (Abbott & Brocht, 2001, cited in Alexander & Madsen, 2009). In their 2009 report, Alexander and Madsen (2009) reported that information technology (IT) use in LTC was shifting toward resident care and clinical support, with less emphasis on administrative functions. However, a more recent report by Alexander and colleagues (2017) indicated that IT use was still greater for administrative functions. The use of IT for resident care and clinical support has been associated with care quality (Alexander & Madsen, 2018; Powell et al., 2020), and greater use of IT for these purposes would likely be beneficial.

Use of Health Assessment Information for Care Planning and Quality Improvement

An obvious focus of efforts to improve care using IT would be on the use of routinely collected health assessment information for resident care planning and quality improvement. Shiells and colleagues (2020) found that the use of electronic patient records could both facilitate (e.g., by allowing access to records at the point of care) and hinder (e.g., due to lack of interoperability) the provision of dementia care. Their study highlighted that systems need to be customizable and that nursing home staff should be consulted during their development, implementation and evaluation.

The standardized assessment system that is likely in widest use in the LTC sector is the interRAI Minimum Data Set (MDS 2.0) assessment system. The MDS/RAI assessment systems were developed by an international consortium of researchers (Hirdes et al., 1999) who support their implementation in many parts of the health care system. Its initial application was in LTC homes in the United States as part of federal efforts to improve nursing home quality; assessments are usually repeated every three months or when warranted by a change in health status (Hawes et al., 1997). The MDS instrument is intended to provide a comprehensive assessment of many aspects of health, physical functioning and service use. In the United States, a revised version—the MDS 3.0—was developed to address perceived shortcomings of the MDS 2.0 (Saliba et al., 2012). Some other centers and jurisdictions are using an updated version of the MDS 2.0 called the interRAI-LTCF (Boorsma et al., 2013).

A study of the implementation of the RAI instrument found moderate to good psychometric properties and positive effects on the care process, although effects on resident function were mixed (Achterberg et al., 1999). Work by Phillips and colleagues (1997) found that the MDS may have resulted in improvements in health outcomes in some areas, but may also have generated trade-offs, due to focusing staff attention on specific subpopulations of residents.

The RAI/MDS assessments contain a number of assessment scales for specific health issues, such as cognition (Morris et al., 1994) or pain (Fries et al., 2001). Clinical assessment protocols (CAPs) are triggered when the assessments indicate risks or problems that warrant further assessment (Morris et al., 1997). Quality indicators for nursing homes have also been derived from these assessment items (Zimmerman et al., 1995). These assessments can thus serve a variety of purposes related to resident care, decision support and quality improvement.

The availability of health data does not mean that these data will necessarily be used for care or decision-making, as has been noted by Prosperi and colleagues (2018) as well as others (Egan et al., 2009; Stolee et al., 2010). A recent exploration of the use of the RAI assessment in Canadian home care settings found that many agencies lacked an understanding of these assessments or were otherwise poorly equipped to use these data for decision-making and quality improvement (Elliott et al., 2020). Also, while recognizing the value of standardized assessment systems, these need to be augmented by clinical observations and judgment (Egan et al., 2009; Lafortune et al., 2017). Effectiveness of systems such as the interRAI assessments is thus dependent on clinician and administrator understanding of the data, support for their application in practice and integration with practice workflows and other available information.

Care Management in LTC using a Cloud-based ITC Tool and the interRAI-LTCF

Researchers in South Korea took advantage of that country's position as one of the most wired countries in the world to implement a cloud-based ICT tool—Systems for Person-centered Elder Care (SPEC)—to support a comprehensive assessment and care management process to enhance the quality of care of LTC residents (Kim et al., 2017). Comprehensive assessment and the measurement of quality indicators were undertaken using the interRAI-LTCF.

In a stepped-wedge cluster randomized trial, Kim and colleagues (2021) found that SPEC significantly improved the quality of care.

A recent Korean study (Kim et al., 2021) provides encouraging evidence of the benefits of the use of the interRAI assessment system as part of a technology-supported integrated care model called Systems for Person-centered Elder Care (SPEC) to improve the quality of care for older residents in LTC homes. In addition to the use of the interRAI-LTCF as a comprehensive geriatric assessment, the intervention included other care planning and coordination protocols along with a cloud-based information system. Using a stepped-wedge cluster randomized trial design, the researchers found that the intervention resulted in significant ($p < 0.0001$) improvements in overall quality of care, with no important adverse events. An implementation study found that ICT support was a key factor in its effectiveness, but recommended development of a more user-friendly system (Choi et al., 2021).

How Could These Technologies Help Agnes?

An interRAI-LTCF type of assessment could help ensure that care plans are based on a comprehensive understanding of Agnes's health and physical function. Consistent monitoring through regular, repeat assessments could alert care staff to unexpected declines in health and to conditions—such as severe pain—which warrant additional management and support. At a facility-wide level, these data can provide indicators that can inform areas for quality improvement, which could benefit Agnes and other residents of the home.

Technologies to Support Communication and Social Interaction

As was noted earlier in this chapter, social isolation and loneliness are common among residents of LTC homes. These challenges were exacerbated in many countries during the COVID-19 pandemic as a result of visitor restrictions, and limitations on opportunities for interaction with other residents (Chu et al., 2021b; Hado & Feinbert, 2020; Veiga-Seijo et al., 2021).

A variety of technologies have been used to support communication and social isolation, and to reduce social isolation and loneliness among LTC residents. Tsai and colleagues (2020) studied the benefits of a smartphone-based videoconferencing program in a sample of older nursing home residents that did not include persons living with dementia. The researchers found that the intervention improved feelings of loneliness and physiological health, but not depression. Wilson and Small (2020) used a concept mapping approach to explore the perspectives of care staff on the use of mobile technology to support communication between care staff and residents; care staff felt that these technologies could assist in overcoming barriers to care, particularly for residents with communication impairments or language barriers. This research group (Wilson et al., 2020a) also conducted a systematic search to identify *communication apps* that could support person-centered care in LTC. The researchers reviewed apps that were intended to support translation or augmentative and alternative communication (AAC) and identified apps most appropriate for use with residents living in LTC. Recommended apps for use in LTC included *Microsoft Translator* and *Google Translate* for translation, and *Communico-Tool* and *GoTalk NOW* for AAC. The researchers highlighted the importance of ongoing reviews of the content and quality of apps, and of including nurses and other care staff in the future development and design of apps for LTC. We would also encourage that residents be involved in these processes.

A recent literature review by Macdonald and colleagues (2021) examined assistive technologies that support social interaction in LTC. The review excluded social robots (discussed in the next section of this chapter). Their review identified technologies which they categorized as low (recognizable to everyone, such as the telephone); medium (involving more electronics, e.g., tablet-based games); and high (requiring Internet, such as videoconferencing and social media platforms). The authors advanced this definition of socially assistive technologies: "user-appropriate devices and tools that enable real-time connectivity to enhance social interaction" (Macdonald et al., 2021, p. 2695). The review found that the benefits of use of these technologies were variable and cost estimates had not been provided. The authors recommend that additional research be done including cost estimates, and that studies should include persons with dementia and diverse backgrounds.

Budak and colleagues (2021) also completed a literature review related to assistive technologies, specifically in relation to their role in addressing loneliness of persons living with dementia. They concluded that assistive technologies had shown promising impacts on social engagement and quality-of-life, but that loneliness was a specific target in only one study. The assistive technologies included multimedia computer systems which present a variety of content (e.g., photographs, music and videos) to support engagement. An example is the CIRCA (Computer Interactive Reminiscence and Conversation Aid) system developed by Astell and colleagues (2010). Budak et al. (2021) recommended more research to determine the potential of assistive technologies to support people living with dementia in LTC.

How Could These Technologies Help Agnes?

Agnes could use a videoconferencing system or other technology to communicate with Lauren and other members of her family who find it difficult to visit in person, or who may be precluded from doing so due to pandemic-related visitor restrictions. As her cognitive impairment progresses, Agnes's communication and engagement could be supported with communication devices and/or multimedia systems that aid in reminiscing and conversation.

Technologies for Entertainment and Recreation

There seems to have been less attention paid to the role technologies might play, as noted earlier, in support of entertainment and recreation opportunities in LTC. That said, there are a variety of entertainment or recreational technologies that could benefit LTC residents.

In a white paper for the AGE-WELL network, Kaufman and colleagues (2020) have explored the potential for digital gameplay to support the health and well-being of older persons. Their review considered online educational games with customized content, gameplay to connect with others and reduce loneliness, games which incorporate physical movement, and intergenerational digital game play. The white paper provides a more in-depth presentation of examples and design considerations, including the following example of the use of exergames for fall prevention:

Exergames for Fall Prevention

In 2013 study (Chao et al., 2013), seven older adults in a Buffalo, New York assisted living residence played Wii Fit games for 30 minutes, twice weekly for eight weeks. Led by two nursing Ph.D. students, they worked in pairs and encouraged each other in sessions that included exercises for range of motion, acrobic conditioning, strength training, balance improvement, and deep breathing and yoga. At the sessions they discussed health education and motivational topics such as reasons to exercise, how to stay safe, and how to keep going. All participants completed the program safely with no falls or injuries. After the program they showed significant improvement in balance, as well as improved mobility, walking distance, and confidence in performing daily activities without falling. Although a large scale study is needed because this sample was very small, these results were consistent and promising. (Kaufman et al., 2020)

"Exergames" integrate movement into game features—examples are the Kinect and Nintendo Wii (Kaufman et al., 2020). Kaufman et al. (2020) cite a 2015 literature review, by Chao et al. (2015), that found that Wii exergames improved physical function, including balance and mobility. Subsequent to the release of the white paper, Chu and colleagues (2022) conducted a systematic review of the literature on exergaming for older adults living in LTC. Their review found evidence of benefits for physical health, but also that more research is needed on the impact of exergaming for cognitive outcomes and quality-of-life.

Intergenerational digital gameplay provides opportunities for younger and older generations to interact but may prove challenging for older adults with physical or cognitive deficits or who are unfamiliar with digital devices (Kaufman et al., 2020). Zhang and Kaufman (2016) reviewed 19 studies of intergenerational activities supported by digital games, of which seven were specifically designed for intergenerational play. They found that a critical factor in successful design of these games was attention to relationship formation. Zhang and Kaufman (2016) recommend that intergenerational games include entertainment components as well as educational components to support knowledge transfer between the generations.

An intriguing gaming system, "Obie" involves interactive games projected on a large physical space, such as a table, which provide a cognitive activity combined with physical movement and social interaction (Goldberg, 2022). The manufacturer, EyeClick, provides reports on its website, Obie For Seniors, Research, of benefits for LTC residents in terms of entertainment and social interaction (Fig. 5.2).

Fig. 5.2 Obie gaming platform in use by older adults during pandemic lockdown. *Source* EyeClick. Photo used with permission

In their review of digital gaming technologies, Kaufman and colleagues (2020) noted the importance of good design as well as technical and social support to help older players learn the game and realize its benefits. As a caveat, little of the research on digital gaming has been done in LTC. Therefore additional research will be needed on how these technologies can be adapted for, and implemented in, LTC homes.

Gaming systems can also be used to monitor health status of LTC residents. For example, Wilkinson and colleagues (2018) describe how ambient activity units that provide entertaining interactions for LTC residents can also be used to collect data on the health status of residents.

Intergenerational digital gaming, if done remotely, could be a way for Agnes to connect with her daughter Lauren and her family. Other digital games might help to alleviate Agnes's boredom and loneliness if appropriate accommodations are made for her arthritis and limited mobility.

Social Robots and Automated Prompting Systems

Social robots have been suggested as a technology that could alleviate social isolation of older persons, particularly during periods of visitor restriction such as during the COVID-19 pandemic (Isabet et al., 2021; Smith et al., 2021). Reports on social robots for older persons have found potential benefits and have also identified challenges warranting further research.

Isabet et al. (2021) completed a literature review of the use of mobile telepresence robots that they describe as "tools equipped with a camera, monitor, microphone, and speaker, with a body on wheels that allows for remote controlled and sometimes autonomous movement aiming to provide easy access to assistance and networking services." Their review found promising results of studies on robot use, as well as technical, ergonomic and ethical challenges that remain to be solved. Ethical issues relate to confidentiality, older adults needing to give consent before using robots and to have control over their use, and equitable access. Co-creation of robotic technology has been identified as a way to identify and manage ethical concerns, such as privacy issues, related to robotic technology (Hardcastle & Ogbogu, 2020; Robillard & Kabacińska, 2020).

Socially assistive robots have been tested for use with older adults, including persons living with cognitive impairment, and in LTC settings (McColl et al., 2013). Robots are being developed to provide assistance during meals or to play games with the older adult (McColl et al., 2013) Researchers have developed approaches for the robots to use emotion-based behavior in providing assistance (Ficocelli et al., 2016). A systematic review by Lu and colleagues (2021) found benefits of companion robots for persons living with dementia in LTC in terms of reduced agitation.

Robots are being developed to emulate a human form; they can also take the form of pets. Scerri et al. (2020) found potential benefits of pet robots for persons living with dementia in LTC, though these were not found to be a positive experience for everyone. The use of robotic seals as therapeutic tools in LTC has been explored. Similarly to the experience reported by Scerri et al. (2020), findings suggested potential benefits of "Paro" as a therapeutic tool, although research has found it is "not for everybody" (Birks et al., 2016). More recently, as systematic review and meta-analysis of the effectiveness of Paro for older adults living with dementia found Paro to have benefits for behavioural and psychological symptoms, medication use and sociability; the review also noted that studies were limited and more rigorous research is needed (Rashid et al., 2023) (Fig. 5.3).

Computerized devices are being developed and tested to assist persons with cognitive impairment to perform activities of daily living; these systems provide automated cueing or prompting to assist in tasks such as dressing (Mahoney et al., 2015) or handwashing (Boger et al., 2006; Hoey et al., 2010; Mihailidis et al., 2008) (Fig. 5.4).

Similar approaches have been used to develop technologies to assist older adults with cognitive impairment to create visual art (Blunsden et al., 2009).

Fig. 5.3 Two LTC residents interact with PARO, a social robot modeled after a baby harp seal. *Source* Paro Robots Inc. Photo used with permission

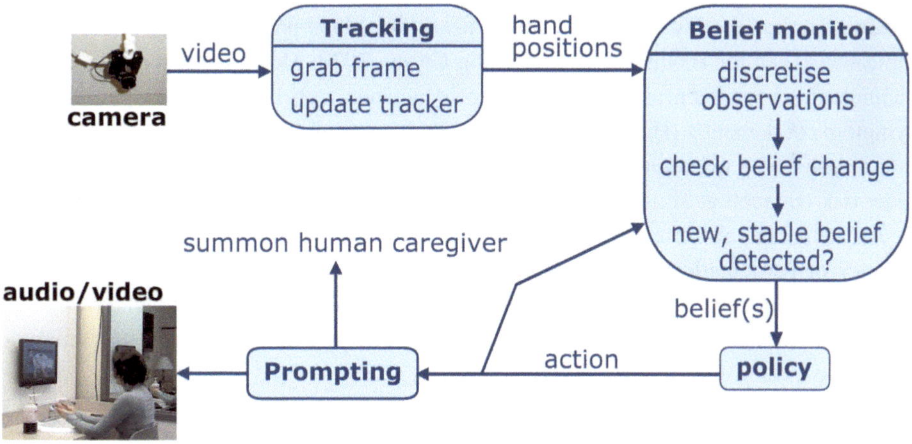

Fig. 5.4 Schematic for a vision-based system to assist a person with dementia to wash their hands. *Source* This image was published in Computer Vision and Image Understanding, vol. 114, Hoey, J., Poupart, P., von Bertoldi, A., Craig, T., Boutilier, C., & Mihailidis A. Automated handwashing assistance for persons with dementia using video and a partially observable Markov decision process. pp. 503–519. Copyright Elsevier (2010). Used with permission

How Could These Technologies Help Agnes?

As Agnes experiences declines in her cognitive function, social robots or automated prompting systems could assist Agnes in her daily activities, such as providing prompts to help her with dressing or at mealtime. Robots could provide her with opportunities for recreation, such as playing a card game. The robots could also provide social support and companionship which would be particularly important given that she lives far away from her family and may also have restricted socialization opportunities for reasons of infection control.

Remote Monitoring Technology, Including Sensors and Wearables

Technologies for remote or passive monitoring have been explored for use in LTC, including for use as a surveillance technology that could be used instead of physical restraints (Te Boekhorst et al., 2013) or for prevention of resident-to-resident abuse (Sifford & Bharucha, 2010). Video surveillance technology has been suggested as a way to monitor quality of care, however this is controversial (Berridge et al., 2019). A more focused use of video technology is for the monitoring of pain through observation of facial expressions; recent research has tested the application of this technology for LTC residents with and without dementia (Rezaei et al., 2021; Taati et al., 2019).

Mobile health solutions, including smartphones or wearable sensors, have been developed to monitor activity levels, gait speed, heart rate, blood pressure, sleep or other health indicators (Data for Healthy Aging Working Group, 2020; Merilahti et al., 2016). Sensor technologies have been used to guide care of persons with behavioral or psychological symptoms of dementia (Husebo et al., 2020). Pressure sensitive mats have been tested as a non-invasive way to monitor bladder voiding (Cohen-McFarlane et al., 2016) or pressure ulcer risk (Bennett et al., 2013).

An advantage of these technologies is that they require little or no technological proficiency or participation by the residents, thus increasing their potential use for persons living with dementia. For the same reason, these technologies also raise potential concerns about consent, confidentiality and privacy (Segura Anaya et al., 2018), and users may not always be aware of these risks (Cilliers, 2020).

These technologies can gather a large amount of data, but these data only become useful if a staff member is seeing the data (either through regular monitoring or by receiving alerts) and is able to make use of the information in a timely fashion. To justify purchase of these technologies, LTC operators would need a business case that they are able to improve the quality of care and quality-of-life of residents without placing additional demands on already limited staff resources.

How Could These Technologies Help Agnes?

Agnes's cognitive impairment, arthritis, limited mobility and incontinence place her at risk for unmanaged pain, pressure ulcers, infections or other health issues. These conditions may be monitored through a periodic assessment, such as with the RAI-LTCF, but events which occur between assessments may not be noticed (or the assessments may not be acted upon). Remote or passive monitoring technologies could provide more continuous monitoring that could enable earlier prevention or treatment.

Seminal Demonstration Project

Social isolation and loneliness are common for persons living in LTC and these concerns were further exacerbated during the COVID-19 pandemic. We propose that a multi-site demonstration project be undertaken to understand how technologies can be used effectively to mitigate the social isolation and loneliness of LTC residents. The technologies to be tested should be identified and implemented through processes of user engagement and co-creation, involving diverse groups of residents, family caregivers, LTC staff and operators. These groups should consider a range of possible technologies, including videoconferencing and other communication technologies such as electronic gaming systems (Kaufman et al., 2020), which can support social interaction and intergenerational connections, social robots and other technologies. The projects should address:

- How these technologies can be funded and procured.
- What space and infrastructure (such as WiFi access) are needed to implement them.
- Implementation with cultural humility and safety.
- How the introduction and sustainability of the technologies will be aided by technical support and training (e.g., one-on-one coaching as proposed by Hoang et al. (2021)).
- Implications for staff time.
- Impacts on social isolation, loneliness and other outcomes such as quality-of-life and stress on family caregivers.

Ideally, this demonstration project will be undertaken in multiple jurisdictions, and include rural, urban, non-profit and for-profit LTC homes.

Concluding Remarks

In this chapter, we have highlighted many challenges facing a LTC system with constrained resources and an increasingly complex and diverse resident population. We have discussed a variety of technologies that might alleviate some of these challenges. We caution that while there are many potential opportunities for technologies to make a positive impact on the quality of care and quality-of-life of LTC residents, they will not in themselves address the long-standing deficiencies of this sector caused by limited funding, inadequate staffing and negative public perceptions.

Technologies should not be considered as the solution to inadequate staffing, or as a significant replacement for direct patient care (Alrassi et al., 2021). Chapman et al. (2019) completed a review of the potential impact of emerging technologies on the LTC workforce and concluded that technology would not fully replace direct care workers, and that robotic solutions were unlikely to mitigate workforce shortages.

Strengthening LTC needs to be considered in the context of the health care system as a whole. It is especially important, as noted by Grabowski (2021), that investments in LTC should not detract from nor devalue home and community services which would be preferred by most older adults, and which also have limited resources (Genet et al., 2011; VanderBent, 2021).

These caveats notwithstanding, the LTC sector and its residents should not be denied the benefit of helpful technologies. Assuming adequate funding can be obtained, the adoption of technological innovations should be supported by streamlined regulatory processes and strengthened procurement mechanisms, including group purchasing arrangements. Basic infrastructure needs to be in place, including high-speed Internet and WiFi access.

The risk-averse LTC sector should be encouraged to work with researchers and innovators in developing and implementing technologies that serve their needs. As part of these efforts, it is essential that introduction of technologies be supported by meaningful engagement of residents, families and staff at all stages of development, implementation and evaluation. These co-creation processes will help to ensure that technological innovations respond to real issues and challenges, that technologies are designed to be used easily by LTC residents and staff, and that ethical issues are identified and managed. Co-creation processes should also engage participation that is reflective of the diverse backgrounds, cultures and health conditions of LTC residents and staff.

References

Abbott, P.A., & Brocht, D. (2001). Challenges for data management in long-term care. In V. K. Saba & K. A. McCormick (Eds.), *Essentials of computers for nurses: Informatics for the new millennium.* (3rd Ed., pp. 357–4). McGraw-Hill.

Achterberg, W., Pot, A. M., van Campen, C., & Ribbe, M. (1999). Het resident assessment instrument (RAI): Een overzicht van internationaal onderzoek naar de psychometrische kwaliteiten en effecten van implementatie in verpleeghuizen [Resident Assessment Instrument (RAI): A review of international research on the psychometric qualities and effects of implementation in nursing homes]. *Tijdschrift Voor Gerontologie En Geriatrie, 30*(6), 264–270. https://doi.org/10.1186/1471-2318-11-20

Alexander, G. L., & Madsen, R. (2009). IT sophistication in nursing homes. *Journal of the American Medical Directors Association, 10*(6), 398–7. PMID: 19560717. PMCID: PMC2791522.

Alexander, G. L., Madsen, R. W., Miller, E. L., Schaumberg, M. K., Holm, A. E., Alexander, R. L., Wise, K. K., Dougherty, M. L., & Gugerty, B. (2017). A national report of nursing home information technology: Year 1 results. *Journal of the American Medical Informatics Association: JAMIA, 24*(1), 67–73. https://doi.org/10.1093/jamia/ocw051

Alexander, G. L., & Madsen, R. (2018). A national report of nursing home quality and information technology: Two-year trends. *Journal of Nursing Care Quality, 33*(3), 200–207. https://doi.org/10.1097/NCQ.0000000000000328

Alrassi, J., Katsufrakis, P. J., & Chandran, L. (2021). Technology can augment, but not replace, critical human skills needed for patient care. *Academic Medicine, 96*(1), 37–43. https://doi.org/10.1097/ACM.0000000000003733

Armstrong, P., & Braedly, S. (Eds.). (2016). *Physical environments for long-term care: Ideas worth sharing*. Canadian Centre for Policy Alternatives. https://policyalternatives.ca/sites/default/files/uploads/publications/National%20Office/2016/11/Physical_Environments_for_Long-term_Care2016.pdf

Astell, A. J., Ellis, M. P., Bernardi, L., Alm, N., Dye, R., Gowans, G., & Campbell, J. (2010). Using a touch screen computer to support relationships between people with dementia and caregivers. *Interacting with Computers, 22*(4), 267–275. https://doi.org/10.1016/j.intcom.2010.03.003

Bennett, S. L., Goubran, R., Rockwood, K., & Knoefel, F. (2013). Monitoring the relief of pressure points for pressure ulcer prevention: A subject dependent approach. In *2013 IEEE international symposium on medical measurements and applications* (pp. 135–138). Gatineau, Quebec, Canada. May 4–5, 2013. https://doi.org/10.1109/MeMeA.2013.6549722

Berridge, C., Halpern, J., & Levy, K. (2019). Cameras on beds: The ethics of surveillance in nursing home rooms. *AJOB Empirical Bioethics, 10*(1), 55–62. https://doi.org/10.1080/23294515.2019.1568320

Berta, W., Laporte, A., Deber, R., Baumann, A., & Gamble, B. (2013). The evolving role of health care aides in the long-term care and home and community care sectors in Canada. *Human Resources for Health, 11*, 25. https://doi.org/10.1186/1478-4491-11-25

Betini, R., Milicic, S., & Lawand, C. (2021). The impact of the COVID-19 pandemic on long-term care in Canada. *Healthcare Quarterly, 24*(3), 13–15.

Birks, M., Bodak, M., Barlas, J., Harwood, J., & Pether, M. (2016). Robotic seals as therapeutic tools in an aged care facility: A qualitative study. *Journal of Aging Research, 2016*, 8569602. https://doi.org/10.1155/2016/8569602

Blunsden, S., Richards, B., Boger, J, Mihailidis, A., Bartindale, T., Jackson, D. G., Olivier, P., & Hoey, J. (2009). Design and prototype of a device to engage cognitively disabled older adults in visual artwork. In *Proceedings of the 2nd international conference on pervasive technologies related to assistive environments*. PETRA, Corfu, Greece, June 9–13, 2009. https://doi.org/10.1145/1579114.1579162

Boger, J., Hoey, J., Poupart, P., Boutilier, C., Fernie, G., & Mihailidis, A. (2006). A planning system based on Markov decision processes to guide people with dementia through activities of daily living. *IEEE Transactions on Information Technology in Biomedicine: A Publication of the IEEE Engineering in Medicine and Biology Society, 10*(2), 323–333. https://doi.org/10.1109/titb.2006.864480

Boorsma, M., Langedijk, E., Frijters, D. H., Nijpels, G., Elfring, T., & van Hout, H. P. (2013). Implementation of geriatric assessment and decision support in residential care homes: Facilitating and impeding factors during initial and maintenance phase. *BMC Health Services Research, 13*, 8. https://doi.org/10.1186/1472-6963-13-8

Boscart, V. M., Sidani, S., Poss, J., Davey, M., d'Avernas, J., Brown, P., Heckman, G., Ploeg, J., & Costa, A. P. (2018). The associations between staffing hours and quality of care indicators in long-term care. *BMC Health Services Research, 18*(1), 750. https://doi.org/10.1186/s12913-018-3552-5

Brown, C. V., & Braun, K. L. (2008). Globalization, women's migration, and the long-term workforce. *The Gerontologist, 48*(1), 16–24. https://doi.org/10.1093/geront/48.1.16

Brown, K. A., Jones, A., Daneman, N., Chan, A. K., Schwartz, K. L., Garber, G. E., Costa, A. P., & Stall, N. M. (2021). Association between nursing home crowding and COVID-19 infection and mortality in Ontario, Canada. *JAMA Internal Medicine, 181*(2), 229–6. https://doi-org.proxy.lib.uwaterloo.ca/, https://doi.org/10.1001/jamainternmed.2020.6466

Budak, K. B., Atefi, G., Hoel, V., Laporte Uribe, F., Meiland, F., Teupen, S., Felding, S. A., & Roes, M. (2021). Can technology impact loneliness in dementia? A scoping review on the role of assistive technologies in delivering psychosocial interventions in long-term care. *Disability and rehabilitation. Assistive Technology*, 1–13. https://doi.org/10.1080/17483107.2021.1984594

Campbell-Enns, H. J., Campbell, M., Rieger, K. L., Thompson, G. N., & Doupe, M. B. (2020). No other safe care option: Nursing home admission as a last resort strategy. *The Gerontologist, 60*(8), 1504–1514. https://doi.org/10.1093/geront/gnaa077

Canadian Institute for Health Information (CIHI). (2018). *Access to palliative care in Canada.* CIHI. https://www.cihi.ca/sites/default/files/document/access-palliative-care-2018-en-web.pdf

Canadian Institute for Health Information (CIHI). (2020a). Pandemic experience in the long-term care sector: How does Canada compare with other countries? CIHI. https://www.cihi.ca/sites/default/files/document/covid-19-rapid-response-long-term-care-snapshot-en.pdf

Canadian Institute for Health Information (CIHI). (2020b). *Profile of residents in residential and hospital-based continuing care, 2019–2020 Quick Stats.* CIHI.

CBC News. (2018, October 21). Butterfly Model of care raising red flags for dementia experts. *CBC News.* https://www.cbc.ca/news/canada/kitchener-waterloo/butterfly-model-of-care-raising-red-flags-for-dementia-experts-1.4870344

Chamberlain, S., Hoben, M., Squires, J., Cummings, G., Norton, P., & Estabrooks, C. (2019). Who is (still) looking after Mom and Dad? Few improvements in care aides' quality-of-work life. *Canadian Journal on Aging/La Revue Canadienne Du Vieillissement, 38*(1), 35–50. https://doi.org/10.1017/S0714980818000338

Chao, Y. Y., Scherer, Y. K., & Montgomery, C. A. (2015). Effects of using Nintendo Wii™ exergames in older adults: A review of the literature. *Journal of Aging and Health, 27*(3), 379–402. https://doi.org/10.1177/0898264314551171

Chao, Y. Y., Scherer, Y. K., Wu, Y. W., Lucke, K. T., & Montgomery, C. A. (2013). The feasibility of an intervention combining self-efficacy theory and Wii Fit exergames in assisted living residents: A pilot study. *Geriatric Nursing, 34*(5), 377–382. https://doi.org/10.1016/j.gerinurse.2013.05.006

Chapman, S., Miller, J., & Spetz, J. (2019). *The impact of emerging technologies on long-term care and the health workforce.* UCSF Health Workforce Research Center on Long-Term Care.

Chaudhury, H., Cooke, H. A., Cowie, H., & Razaghi, L. (2018). The influence of the physical environment on residents with dementia in long-term care settings: A review of the empirical literature. *The Gerontologist, 58*(5), e325-37. https://doi.org/10.1093/geront/gnw259

Cherry, B. J., Ford, E. W., & Peterson, L. T. (2011). Experiences with electronic health records: Early adopters in long-term care facilities. *Health Care Management Review, 36*(3), 265–274. https://doi.org/10.1097/HMR.0b013e31820e110f

Choi, H., Jung, Y. I., & Kim, H. (2021). Implementation fidelity of the systems for person-centered elder care (SPEC): A process evaluation study. *Implementation Science: IS, 16*(1), 52. https://doi.org/10.1186/s13012-021-01113-3

Chu, C. H., Ronquillo, C., Khan, S., Hung, L., & Boscart, V. (2021). Technology recommendations to support person-centered care in long-term care homes during the COVID-19 pandemic and beyond. *Journal of Aging & Social Policy, 33*(4–5), 539–354. https://doi.org/10.1080/08959420.2021.1927620

Chu, C. H., Wang, J., Fukui, C., Staudacher, S., A Wachholz, P., & Wu, B. (2021b). The impact of COVID-19 on social isolation in long-term care homes: Perspectives of policies and strategies from six countries. *Journal of Aging & Social Policy*, *33*(4–5), 459–73. https://doi.org/10.1080/08959420.2021.1924346

Chu, C. H., Quan, A., Souter, A., Krisnagopal, A., & Biss, R. K. (2022). Effects of exergaming on physical and cognitive outcomes of older adults living in long-term care homes: A systematic review. *Gerontology, 68*(9), 1–17. https://doi.org/10.1159/000521832

Cilliers, L. (2020). Wearable devices in healthcare: Privacy and information security issues. *Health Information Management: Journal of the Health Information Management Association of Australia, 49*(2–3), 150–156. https://doi.org/10.1177/1833358319851684

Clemens, S., Wodchis, W., McGilton, K., McGrail, K., & McMahon, M. (2021). The relationship between quality and staffing in long-term care: A systematic review of the literature 2008–2020. *International Journal of Nursing Studies, 122*, 104036. https://doi.org/10.1016/j.ijnurstu.2021.104036

Cohen-McFarlane, M., Green, J. R., Knoefel, F., & Goubran, R. (2016). Smart monitoring of fluid intake and bladder voiding using pressure sensitive mats. In *2016 38th annual international conference of the IEEE engineering in medicine and biology society* (pp. 4921–4924) [Conference presentation]. https://doi.org/10.1109/EMBC.2016.7591831

Cook, E. J., Randhawa, G., Sharp, C., Ali, N., Guppy, A., Barton, G., Bateman, A., & Crawford-White, J. (2016). Exploring the factors that influence the decision to adopt and engage with an integrated assistive telehealth and telecare service in Cambridgeshire, UK: A nested qualitative study of patient 'users' and 'non-users'. *BMC Health Services Research, 16*, Article no. 137. https://doi.org/10.1186/s12913-016-1379-5

Cooney, A. (2011). 'Finding home': A grounded theory on how older people 'find home' in long-term care settings. *International Journal of Older People Nursing, 7*(3), 188–199. https://doi.org/10.1111/j.1748-3743.2011.00278.x

Corazzini, K., Twersky, J., White, H. K., Buhr, G. T., McConnell, E. S., Weiner, M., & Colón-Emeric, C. S. (2015). Implementing culture change in nursing homes: An adaptive leadership framework. *The Gerontologist, 55*(4), 616–627. https://doi.org/10.1093/geront/gnt170

Data for Healthy Aging Working Group. (2020). *AGE-WELL position paper on mobile technology and data-informed approaches for healthy aging and aging-in-place.* University of Calgary.

Declercq, A., de Stampa, M., Geffen, L., Heckman, G., Hirdes, J., Finne-Soveri, H., Lum, T., Millar, N., Morris, J.N., Onder, G., Szczerbinska, K., Topinkova, E., & van Hout, H. (2020). *Why, in almost all countries, was residential care for older people so badly affected by COVID-19?* OSE Working Paper Series, Opinion paper no. 23. European Social Observatory. https://www.interrai.co.nz/assets/Declercq_Et_Al_2020_OpinionPaper23-1-002.pdf

Dyer, S. M., Valeri, M., Arora, N., Ross, T., Winsall, M., Tilden, D., & Crotty, M. (2019). *Review of international systems for long-term care of older people.* Flinders University, Australia. https://agedcare.royalcommission.gov.au/publications/Documents/research-paper-2-review-international-systems-long-term-care.pdf

Dykgraaf, S. H., Matenge, S., Desborough, J., Sturgiss, E., Dut, G., Roberts, L., McMillan, A., & Kidd, M. (2021). Protecting nursing homes and long-term care facilities from COVID-19: A rapid review of the evidence. *JAMDA, 22*(10), 1969–1988. https://doi.org/10.1016/j.jamda.2021.07.027

Egan, M., Wells, J., Byrne, K., Jaglal, S., Stolee, P., Chesworth, B. M., & Hillier, L. M. (2009). The process of decision-making in home-care case management: Implications for the introduction of universal assessment and information technology. *Health & Social Care in the Community, 17*(4), 371–378. https://doi.org/10.1111/j.1365-2524.2008.00835.x

Elliott, J., Gordon, A., Tong, C. E., & Stolee, P. (2020). We've got the home care data, what do we do with it? Understanding data use in decision making and quality improvement. *BMC Health Services Research, 20*(1), 251. https://doi.org/10.1186/s12913-020-5018-9

Emodi, B. (1977). *A history of long-term care facilities for the elderly in Canada.* Canada Mortgage and Housing Corporation. Series title: Nursing and hostels design guidelines study. Working paper no. 4. https://publications.gc.ca/collections/collection_2016/schl-cmhc/NH18-25-4-1977-eng.pdf

Employment and Social Development Canada. (2018). *Social isolation of seniors: A focus on new immigrant and refugee seniors in Canada.* Her Majesty the Queen in Right of Canada.

Estabrooks, C.A., Straus, S., Flood, C.M., Keefe, J., Armstrong, P., Donner, G., Boscart, V., Ducharme, F., Silvius, J., & Wolfson, M. (2020). *Restoring trust: COVID-19 and the future of long-term care.* Royal Society of Canada. https://rsc-src.ca/sites/default/files/LTC%20PB%20%2B%20ES_EN.pdf

Esteves, M., Esteves, M., Abelha, A., & Machado, J. (2019). A proof of concept of a mobile health application to support professionals in a Portuguese nursing home. *Sensors (Basel, Switzerland), 19*(18), 3951. https://doi.org/10.3390/s19183951

Fasullo, K., McIntosh, E., Buchholz, S. W., Ruppar, T., & Ailey, S. (2021). LGBTQ older adults in long-term care settings: An integrative review to inform best practices. *Clinical Gerontologist,* 1–16. https://doi.org/10.1080/07317115.2021.1947428

Ficocelli, M., Terao, J., & Nejat, G. (2016). Promoting interactions between humans and robots using robotic emotional behavior. *IEEE Transactions on Cybernetics, 46*(12), 2911–2923. https://doi.org/10.1109/TCYB.2015.2492999

Forbes, W. F., Jackson, J. A., & Kraus, A. S. (1987). *Institutionalization of the elderly in Canada.* Butterworths.

Freedman, A., & Nicolle, J. (2020). Social isolation and loneliness: The new geriatric giants: Approach for primary care. *Canadian Family Physician/Medecin de famille canadien, 66*(3), 176–82. PMCID: PMC8302356.

Fries, B. E., Simon, S. E., Morris, J. N., Flodstrom, C., & Bookstein, F. L. (2001). Pain in U.S. nursing homes: Validating a pain scale for the minimum data set. *The Gerontologist, 41*(2), 173–79. https://doi.org/10.1093/geront/41.2.173

Gahwi, L., & Walton-Roberts, M. (2020, June 23). Migrant care labour and the COVID-19 long-term care crisis: How did we get here? *Balsillie Papers. 2*(2). https://balsilliepapers.ca/wp-content/uploads/2020/12/Balsillie-Paper-Gahwi-and-Walton-Roberts-RT.pdf

Gauvin, F. P., Wilson, M. G., DeMaio, P., Alam, S., Drakos, A., & Lavis, J. N. (2021). *Evidence brief: Identifying and harnessing the potential of technology in long-term care settings in Canada.* McMaster Health Forum.

Genet, N., Boerma, W. G., Kringos, D. S., Bouman, A., Francke, A. L., Fagerström, C., Melchiorre, M. G., Greco, C., & Devillé, W. (2011). Home care in Europe: A systematic literature review. *BMC Health Services Research, 11,* Article no. 207. https://doi.org/10.1186/1472-6963-11-207

Giwa, S., Mullings, D. V., & Karki, K. K. (2020). Virtual social work care with older black adults: A culturally relevant technology-based intervention to reduce social isolation and loneliness in a time of pandemic. *Journal of Gerontological Social Work, 63*(6–7), 679–681. https://doi.org/10.1080/01634372.2020.1800885

Goldberg, E. (2022). *Obie training through gaming: Cognitive enhancement and digital technologies.* https://seniors-assets.s3.amazonaws.com/Documents/OBIE+MAIN+Training+Through+Gaming+Aug+2020+%5BEN%5D.pdf

Grabowski, D. C. (2021). The future of long-term care requires investment in both facility- and home-based services. *Nature Aging, 1*, 10–11. https://doi.org/10.1038/s43587-020-00018-y

Hado, E., & Feinbert, L. F. (2020). Amid the COVID-19 pandemic, meaningful communication between family caregivers and residents of long-term care facilities is imperative. *Journal of Aging & Social Policy, 32*(4–5), 410–415. https://doi.org/10.1080/08959420.2020.1765684

Hardcastle, L., & Ogbogu, U. (2020). Virtual care: Enhancing access or harming care? *Healthcare Management Forum, 33*(6), 288–292. https://doi.org/10.1177/0840470420938818

Haslam, D. (2013). Do we just pay lip service to patient-centred care? *The Practitioner, 257*(1758), 35. PMID: 23577520.

Hawes, C., Mor, V., Phillips, C. D., Fries, B. E., Morris, J. N., Steele-Friedlob, E., Greene, A. M., & Nennstiel, M. (1997). The OBRA-87 nursing home regulations and implementation of the Resident Assessment Instrument: Effects on process quality. *Journal of the American Geriatrics Society, 45*(8), 977–985. https://doi.org/10.1111/j.1532-5415.1997.tb02970.x

Hawes, C. (2003). Elder abuse in residential long-term care settings: What is known and what information is needed? In J. R. Bonnie and R. B. Wallace (Eds.), *Elder mistreatment: Abuse, neglect, and exploitation in an aging America* (pp. 446–00). National Academies Press (US).

Heckman, G. A., Kay, K., Morrison, A., Grabowski, D. C., Hirdes, J. P., Mor, V., Shaw, G., Benjamin, S., Boscart, V. M., Costa, A. P., Declercq, A., Geffen, L., Sang Lum, T. Y., Moser, A., Onder, G., & van Hout, H. (2021). Proceedings from an International Virtual Townhall: Reflecting on the COVID-19 Pandemic: Themes from long-term care. *Journal of the American Medical Directors Association, 22*(6), 1128–1132. https://doi.org/10.1016/j.jamda.2021.03.029

Hirdes, J. P., Fries, B. E., Morris, J. N., Steel, K., Mor, V., Frijters, D., LaBine, S., Schalm, C., Stones, M. J., Teare, G., Smith, T., Marhaba, M., Pérez, E., & Jónsson, P. (1999). Integrated health information systems based on the RAI/MDS series of instruments. *Healthcare Management Forum, 12*(4), 30–40. https://doi.org/10.1016/S0840-4704(10)60164-0

Hoang, P., Whaley, C., Thompson, K., Ho, V., Rehman, U., Boluk, K., & Grindrod, K. A. (2021). Evaluation of an intergenerational and technological intervention for loneliness: Protocol for a feasibility randomized controlled trial. *JMIR Research Protocols, 10*(2), e23767. https://doi.org/10.2196/23767

Hodgkin, P., & Taylor, J. (2013). Power to the people: What will bring about the patient centred revolution? *BMJ, 347*, f6701. https://doi.org/10.1136/bmj.f6701

Hoey, J., Poupart, P., von Bertoldi, A., Craig, T., Boutilier, C., & Mihailidis, A. (2010). Automated handwashing assistance for persons with dementia using video and a partially observable Markov decision process. *Computer Vision and Image Understanding, 114*, 503–519. https://doi.org/10.1016/j.cviu.2009.06.008

Hsu, A. T., Lane, N., Sinha, S. K., Dunning, J., Dhuper, M., Kahiel, Z., & Sveistrup, H. (2020). Impact of COVID-19 on residents of Canada's long-term care homes—Ongoing challenges and policy response. LTCovid.org. *International Long-Term Care Policy Network, 17*, 1–18. https://ltccovid.org/wp-content/uploads/2020/06/LTCcovid-country-reports_Canada_June-4-2020.pdf

Husebo, B. S., Heintz, H. L., Berge, L. I., Owoyemi, P., Rahman, A. T., & Vahia, I. V. (2020). Sensing technology to monitor behavioral and psychological symptoms and to assess treatment response in people with dementia. A systematic review. *Frontiers in Pharmacology, 10*, 1699. https://doi.org/10.3389/fphar.2019.01699

Isabet, B., Pino, M., Lewis, M., Benveniste, S., & Rigaud, A.-S. (2021). Social telepresence robots: A narrative review of experiments involving older adults before and during the COVID-19 pandemic. *International Journal of Environmental Research and Public Health, 18*(7), 3597. https://doi.org/10.3390/ijerph18073597

Jain, B., Cheong, E., Bugeja, L., & Ibrahim, J. (2019). International transferability of research evidence in residential long-term care: A comparative analysis of aged care systems in 7 nations. *Journal of the American Medical Directors Association, 20*(12), 1558–1565. https://doi.org/10.1016/j.jamda.2019.07.017

Kaufman, D., Sauvé, L., & Ireland, A. (2020). *Playful aging: Digital games for older adults. A white paper by the AGE-WELL 4.2 Project.* Simon Fraser University and Université TÉLUQ, February. https://agewell-nce.ca/archives/9316

Kim, H., Park, Y. H., Jung, Y. I., Choi, H., Lee, S., Kim, G. S., Yang, D. W., Paik, M. C., & Lee, T. J. (2017). Evaluation of a technology-enhanced integrated care model for frail older persons: Protocol of the SPEC study, a stepped-wedge cluster randomized trial in nursing homes. *BMC Geriatrics, 17*(1), 88. https://doi.org/10.1186/s12877-017-0459-7

Kim, H., Jung, Y. I., Kim, G. S., Choi, H., & Park, Y. H. (2021). Effectiveness of a technology-enhanced integrated care model for frail older people: A stepped-wedge cluster randomized trial in nursing homes. *The Gerontologist, 61*(3), 460–469. https://doi.org/10.1093/geront/gnaa090

Koehn, S., Baumbusch, J., Reid, R. C., & Li, N. (2018). 'It's like chicken talking to ducks' and other challenges to families of Chinese immigrant older adults in long-term residential care. *Journal of Family Nursing, 24*(2), 156–183. https://doi.org/10.1177/1074840718774068

Kuluski, K., Williams, A. P., Laporte, A., & Berta, W. (2012). The role of community-based care capacity in shaping risk of long-term care facility placement. *Healthcare Policy—Politiques de sante, 8*(1), 92–105 (PMID: 23968606. PMCID: PMC3430157).

Lafortune, C., Elliott, J., Egan, M. Y., & Stolee, P. (2017). The rest of the story: A qualitative study of complementing standardized assessment data with informal interviews with older patients and families. *The Patient, 10*(2), 215–224. https://doi.org/10.1007/s40271-016-0193-9

Levitt, D. (2021). Transforming the hospital-style nursing home to a centre for living. *Healthcare Management Forum. 35*(1), 35–8. https://doi.org/10.1177/08404704211045053

Li, J., Xu, X., Pham, T. P., Theng, Y. L., Katajapuu, N., & Luimula, M. (2017). Exergames designed for older adults: A pilot evaluation on psychosocial well-being. *Games for Health Journal, 6*(6), 371–378. https://doi.org/10.1089/g4h.2017.0072

Lu, L. C., Lan, S. H., Hsieh, Y. P., Lin, L. Y., Lan, S. J., & Chen, J. C. (2021). Effectiveness of companion robot care for dementia: A systematic review and meta-analysis. *Innovation in Aging, 5*(2), igab013. https://doi.org/10.1093/geroni/igab013

MacDonald, B-J., Wolfson, M., & Hirdes, J. P. (2019). *The future co$t of long-term care in Canada.* National Institute on Ageing, Ryerson. https://static1.squarespace.com/static/5c2fa7b03917eed9b5a436d8/t/5dbadf6ce6598c340ee6978f/1572527988847/The+Future+Cost+of+Long-Term+Care+in+Canada.pdf

MacDonald, M., Yu, Z., Weeks, L. E., Moody, E., Wilson, B., Almukhaini, S., Martin-Misener, R., Sim, M., Jefferies, K., Iduye, D., Neeb, D., & McKibbon, S. (2021). Assistive technologies that support social interaction in long-term care homes: A scoping review. *JBI Evidence Synthesis, 19*(10), 2695–2738. https://doi.org/10.11124/JBIES-20-00264

Mahoney, D. F., Burleson, W., Lozano, C., Ravishankar, V., & Mahoney, E. L. (2015). Prototype development of a responsive emotive sensing system (DRESS) to aid older persons with dementia to dress independently. *Gerontechnology: International Journal on the Fundamental Aspects of Technology to Serve the Ageing Society, 13*(3), 345–58. https://doi.org/10.4017/gt.2015.13.3.005.00

Marquardt, G., Bueter, K., & Motzek, T. (2014). Impact of the design of the built environment on people with dementia: An evidence-based review. *HERD: Health Environments Research & Design Journal, 8*(1), 127–57. https://doi.org/10.1177/193758671400800111

Martinsson, I., Edberg, A.-K., & Janlöv, A.-C. (2013). Growing old in a foreign context—Older immigrants' experience of everyday life in residential care facilities. *Vård i Norden., 33*(3), 34–38. https://doi.org/10.1177/010740831303300308

Marshall, E. G., Clarke, B., Burge, F., Varatharasan, N., Archibald, G., & Andrew, M. K. (2016). Improving continuity of care reduces emergency department visits by long-term care residents. *Journal of the American Board of Family Medicine, 29*(2), 201–208. https://doi.org/10.3122/jabfm.2016.12.150309

McColl, D., & Nejat, G. (2013). Meal-time with a socially assistive robot and older adults in a long-term care facility. *Journal of Human-Robot Interaction, 2*(1), 152–171. https://doi.org/10.5898/JHRI.2.1.McColl

McNeil, H., Elliott, J., Huson, K., Ashbourne, J., Heckman, G., Walker, J., & Stolee, P. (2016). Engaging older adults in healthcare research and planning: A realist synthesis. *Research Involvement and Engagement, 2*(10). https://doi.org/10.1186/s40900-016-0022-2

McNeil, H., McMurray, J., Byrne, K., Grindrod, K., McKinnon, A., & Stolee, P. T. (2017). Engaging older adults and their caregivers in innovation ecosystems for health and aging. *Innovation in Aging, 1*(Suppl 1), 212. https://doi.org/10.1093/geroni/igx004.799

Merilahti, J., Viramo, P., & Korhonen, I. (2016). Wearable monitoring of physical functioning and disability changes, circadian rhythms and sleep patterns in nursing home residents. *IEEE Journal of Biomedical and Health Informatics, 20*(3), 856–864. https://doi.org/10.1109/JBHI.2015.2420680

Mihailidis, A., Boger, J. N., Craig, T., & Hoey, J. (2008). The COACH prompting system to assist older adults with dementia through handwashing: An efficacy study. *BMC Geriatrics, 8*(28). https://doi.org/10.1186/1471-2318-8-28

Ministry of Long-Term Care (2020). *Long-term care staffing study. Long-term care staffing study advisory group. July 30.* https://files.ontario.ca/mltc-long-term-care-staffing-study-en-2020-07-31.pdf

Morris, J., Fries, B. E., Mehr, D. R., Hawes, C., Phillips, C. D., Mor, V., & Lipsitz, L. A. (1994). MDS cognitive performance scale. *Journals of Gerontology: Medical Sciences, 49*(4), M174–M182. PMID: 8014392. https://doi.org/10.1093/geronj/49.4.m174

Morris, J. N., Fries, B. E., Steel, K., Ikegami, N., Bernabei, R., Carpenter, G. I., Gilgen, R., Hirdes, J. P., & Topinkova, E. (1997). Comprehensive clinical assessment in community setting: Applicability of the MDS-HC. *Journal of the American Geriatrics Society, 45*(8), 1017–1024. https://doi.org/10.1111/j.1532-5415.1997.tb02975.x

Mullings, D. V., & Gien, L. (2013). Culturally competent (appropriate) health and long-term care services for older immigrants in a small urban center of Newfoundland. *International Journal of Humanities and Social Science Invention, 2*(9), 81–89.

National Institute on Aging. (2021). *Aging in place: Growing older at home.* U.S. Department of Health & Human Services. https://www.nia.nih.gov/health/aging-place-growing-older-home

Neal, I., du Toit, S., & Lovarini, M. (2020). The use of technology to promote meaningful engagement for adults with dementia in residential aged care: A scoping review. *International Psychogeriatrics, 32*(8), 913–935. https://doi.org/10.1017/S1041610219001388

Office of the Parliamentary Budget Officer. (2021). *Cost estimate for motion 77: Improvements to long-term care.* Office of the Parliamentary Budget Officer. https://publications.gc.ca/collections/collection_2021/dpb-pbo/YN5-229-2021-eng.pdf

Ontario Ministry of Long-Term Care. (2024). *Explore your care options.* https://www.ontario.ca/page/explore-your-care-options#section-3

Organisation for Economic Co-operation and Development (OECD). (2017). Long-term care beds in institutions and hospitals. In *OECD health at a glance 2017: OECD indicators*. OECD Publishing. https://doi.org/10.1787/health_glance-2017-en
Organisation for Economic Co-operation and Development (OECD). (2022). OECD.Stat. *Long-term care resources and utilisation*. https://stats.oecd.org/index.aspx?DataSetCode=HEALTH_LTCR/
Parker, V. A., & Geron, S. M. (2007). Cultural competence in nursing homes. *Gerontology & Geriatrics Education, 28*(2), 37–54. https://doi.org/10.1300/J021v28n02_04
Parmenter, G., Cruickshanck, M., & Hussain, R. (2012). The social lives of rural Australian nursing home residents. *Ageing & Society, 32*(2), 329–353. https://doi.org/10.1017/S0144686X11000304
Phillips, C. D., Morris, J. N., Hawes, C., Fries, B. E., Mor, V., Nennstiel, M., & Iannacchione, V. (1997). Association of the Resident Assessment Instrument (RAI) with changes in function, cognition, and psychosocial status. *Journal of the American Geriatrics Society, 45*(8), 986–993. https://doi.org/10.1111/j.1532-5415.1997.tb02971.x
Picard, A. (2021). *Neglected no more: The urgent need to improve the lives of Canada's elders in the wake of a pandemic.* Penguin Random House Canada Limited.
Poss, J., Murphy, K. M., Woodbury, M. G., Orsted, H., Stevenson, K., Williams, G., MacAlpine, S., Curtin-Telegdi, N., & Hirdes, J. P. (2010). Development of the interRAI Pressure Ulcer Risk Scale (PURS) for use in long-term care and home care settings. *BMC Geriatrics, 10*, Article no. 67. https://doi.org/10.1186/1471-2318-10-67
Powell, K. R., Deroche, C. B., Carnahan, E. J., & Alexander, G. L. (2020). Exploring resident care information technology use and nursing home quality. *Journal of Gerontological Nursing, 46*(4), 15–20. https://doi.org/10.3928/00989134-20200303-02
Power, G. A., & Carson, J. (2021). The promise of transformed long-term care homes: Evidence from the pandemic. *Healthcare Management Forum. 35*(1), 25–8. https://doi.org/10.1177/08404704211037794
Prosperi, M., Min, J. S., Bian, J., & Modave, F. (2018). Big data hurdles in precision medicine and precision public health. *BMC Medical Informatics and Decision Making, 18*, Article no. 139. https://doi.org/10.1186/s12911-018-0719-2
Rashid, N. L. A., Leow, Y., Klainin-Yobas, P., Itoh, S., & Wu, V. X. (2023). The effectiveness of a therapeutic robot, 'Paro', on behavioural and psychological symptoms, medication use, total sleep time and sociability in older adults with dementia: A systematic review and meta-analysis. *International Journal of Nursing Studies, 145*, 104530. https://doi.org/10.1016/j.ijnurstu.2023.104530
Rezaei, S., Moturu, A., Zhao, S., Prkachin, K. M., Hadjistavropoulos, T., & Taati, B. (2021). Unobtrusive pain monitoring in older adults with dementia using pairwise and contrastive training. *IEEE Journal of Biomedical and Health Informatics, 25*(5), 1450–1462. https://doi.org/10.1109/JBHI.2020.3045743
Robillard, J. M., & Kabacińska, K. (2020). Realizing the potential of robotics for aged care through co-creation. *Journal of Alzheimer's Disease: JAD, 76*(2), 461–466. https://doi.org/10.3233/JAD-200214
Roblin, B., Deber, R., Kuluski, K., & Silver, M. (2019). Ontario's retirement homes and long-term care homes: A comparison of care services and funding regimes. *Canadian Journal on Aging / La Revue Canadienne Du Vieillissement, 38*(2), 155–167. https://doi.org/10.1017/S0714980818000569
Saliba, D., Jones, M., Streim, J., Ouslander, J., Berlowitz, D., & Buchanan, J. (2012). Overview of significant changes in the minimum data set for nursing homes version 3.0. *Journal of the American Medical Directors Association, 13*(7), 595–601. https://doi.org/10.1016/j.jamda.2012.06.001

Scerri, A., Sammut, R., & Scerri, C. (2020). Formal caregivers' perceptions and experiences of using pet robots for persons living with dementia in long-term care: A meta-ethnography. *Journal of Advanced Nursing, 77*(1), 83–97. https://doi.org/10.1111/jan.14581

Segura Anaya, L. H., Alsadoon, A., Costadopoulos, N., & Prasad, P. W. C. (2018). Ethical implications of user perceptions of wearable devices. *Science and Engineering Ethics, 24*(1), 1–28. https://doi.org/10.1007/s11948-017-9872-8

Sepulveda, E. R., Stall, N. M., & Sinha, S. K. (2020). A comparison of COVID-19 mortality rates among long-term care residents in 12 OECD countries. *Journal of the American Medical Directors Association, 21*(11), P1572–1574 E3. https://doi.org/10.1016/j.jamda.2020.08.039

Shiells, K., Diaz Baquero, A. A., Štěpánková, O., & Holmerová, I. (2020). Staff perspectives on the usability of electronic patient records for planning and delivering dementia care in nursing homes: A multiple case study. *BMC Medical Informatics and Decision Making, 20*, Article no. 159. https://doi.org/10.1186/s12911-020-01160-8

Sifford, K. S., & Bharucha, A. (2010). Benefits and challenges of electronic surveillance in nursing home research. *Research in Gerontological Nursing, 3*(1), 5–10. https://doi.org/10.3928/19404921-20090706-01

Sloane, P. D., Yearby, R., Konetzka, R. T., Li, Y., Espinoza, R., & Zimmerman, S. (2021). Addressing systemic racism in nursing homes: A time for action. *Journal of the American Medical Directors Association, 22*(4), 886–892. https://doi.org/10.1016/j.jamda.2021.02.023

Smith, C., Gregorio, M., & Hung, L. (2021). Facilitators and barriers to using telepresence robots in aged care settings: A scoping review protocol. *British Medical Journal Open, 11*(9), e051769. https://doi.org/10.1136/bmjopen-2021-051769

Spurgaitis, K. (2018). Emotions: More of a focus in dementia care. *Rehab & Community Care Medicine.* https://www.rehabmagazine.ca/uncategorized/emotions-more-of-a-focus-in-dementia-care/

Stinchcombe, A., Smallbone, J., Wilson, K., & Kortes-Miller, K. (2017). Healthcare and end-of-life needs of lesbian, gay, bisexual, and transgender (LGBT) older adults: A scoping review. *Geriatrics, 2*(1), 13. https://doi.org/10.3390/geriatrics2010013

Stolee, P., Steeves, B., Glenny, C., & Filsinger, S. (2010). The use of electronic health information systems in home care: Facilitators and barriers. *Home Healthcare Nurse, 28*(3), 167–181. https://doi.org/10.1097/01.NHH.0000369769.32246.92

Stolee, P., Zaza, C., & Sharratt, M. T. (2014). Later life care planning conversations for older adults and families. *Journal of Applied Gerontology: The Official Journal of the Southern Gerontological Society, 33*(6), 710–736. https://doi.org/10.1177/0733464813500712

Stone, R. (2019). Strengthening care staff in the long-term care sector: International research perspectives. *Journal of Applied Gerontology: The Official Journal of the Southern Gerontological Society, 38*(11), 1515–1517. https://doi.org/10.1177/0733464818198699

Sue Cragg Consulting & the CLRI Program (2017). *Supporting Indigenous culture in long-term care homes: Needs assessment and ideas 2017–18.* Ontario Centres for Learning, Research & Innovation in Long-Term Care. https://clri-ltc.ca/files/2018/10/Supporting-Indigenous-Culture-in-LTCH-Report-Final-March-31-2017.pdf

Taati, B., Zhao, Z., Ashraf, A. B., Asgarian, A., Browne, M. E., Prkachin, K. M., Mihailidis, A., & Hadjistavropoulos, T. (2019). Algorithmic bias in clinical populations—Evaluating and improving facial analysis technology in older adults with dementi *IEEE Access, 7*, 25527–25534, https://doi.org/10.1109/ACCESS.2019.2900022

Te Boekhorst, S., Depla, M. F. I. A., Francke, A. L., Twisk, J. W., Zwijsen, S. A., & Hertogh, C. M. (2013). Quality of life of nursing-home residents with dementia subject to surveillance technology versus physical restraints: An explorative study. *International Journal of Geriatric Psychiatry, 28*(4), 356–363. https://doi.org/10.1002/gps.3831

Tsai, H. H., Cheng, C. Y., Shieh, W. Y., & Chang, Y. C. (2020). Effects of a smartphone-based videoconferencing program for older nursing home residents on depression, loneliness, and quality of life: A quasi-experimental study. *BMC Geriatrics, 20*(1), 27. https://doi.org/10.1186/s12877-020-1426-2

Twigg, D. E., Whitehead, L., Doleman, G., & El-Zaemey, S. (2021). The impact of nurse staffing methodologies on nurse and patient outcomes: A systematic review. *Journal of Advanced Nursing, 77*(12), 4599–4611. https://doi.org/10.1111/jan.1490

VanderBent, S. (2021, March 12). Invest in home care to avoid creating a second crisis in senior care. *The Standard.* https://www.stcatharinesstandard.ca/opinion/contributors/2021/03/12/invest-in-home-care-to-avoid-creating-a-second-crisis-in-senior-care.html

Veiga-Seijo, R., del Carmen Miranda-Duro, M., & Veiga-Seijo, S. (2021). Strategies and actions to enable meaningful connections in nursing homes during the COVID-19: A scoping review. *Clinical Gerontologist, 45*(1), 20–30. https://doi.org/10.1080/07317115.2021.1937424

Versaevel, C. (2018). *Implementing the butterfly household model of care in Canada: Lessons learned to date.* BC Care Providers Association. https://bccare.ca/2018/01/implementing-butterfly-household-model-care-canada-lessons-learned-date/

Vogel, L. (2018). Pilot project delivers dementia care that feels like home. *CMAJ, 190*(23), E729. https://doi.org/10.1503/cmaj.109-5610

Wagner, L. M., & Rust, T. B. (2008). *Safety in long-term care settings: Broadening the patient safety agenda to include long-term care services.* Canadian Patient Safety Institute. https://iccer.ca/pdf/ccne/SafetyinLTCsettings_2008.pdf

Wang, T., Wang, Y., & Moczygemba, J. (2014). Organizational factors influencing health information technology adoption in long-term-care facilities. *The Health Care Manager, 33*(1), 30–37. https://doi.org/10.1097/01.HCM.0000440624.85255.8e

Welsh, M. (2021). *Happily ever older: Revolutionary approaches to long-term care.* ECW Press.

Wilkinson, A., Tong, T., Zare, A., Kanik, M., & Chignell, M. (2018). Monitoring health status in long term care through the use of ambient technologies and serious games. *IEEE Journal of Biomedical and Health Informatics, 22*(6), 1807–1813. https://doi.org/10.1109/JBHI.2018.2864686

Williams, A. P., Challis, D., Deber, R., Watkins, J., Kuluski, K., Lum, J. M., & Daub, S. (2009). Balancing institutional and community-based care: Why some older persons can age successfully at home while others require residential long-term care. *Healthcare Quarterly, 12*(2), 95–105. https://doi.org/10.12927/hcq.2009.3974

Wilson, K., Kortes-Miller, K., & Stinchcombe, A. (2018). Staying out of the closet: LGBT older adults' hopes and fears in considering end-of-life. *Canadian Journal on Aging, 37*(1), 22–31. https://doi.org/10.10107/S0714980817000514

Wilson, R., Cochrane, D., Mihailidis, A., & Small, J. (2020). Mobile apps to support caregiver-resident communication in long-term care: Systematic search and content analysis. *JMIR Aging, 3*(1), e17136. https://doi.org/10.2196/17136

Wilson, R., & Small, J. (2020). Care staff perspectives on using mobile technology to support communication in long-term care: Mixed methods study. *JMIR Nursing, 3*(1), e21881. https://doi.org/10.2196/21881

Wilson, M. G., Gauvin, F. P., DeMaio, P., Alam, S., Drakos, A., Soueidan, S., Costa, A., Reid, R., Simeonov, D., Sixsmith, A., Sveistrup, H., & Lavis, J. N. (2022). Enhancing the use of technology in the long-term care sector in Canada: Insights from citizen panels and a national stakeholder dialogue. *Healthcare Management Forum, 35*(5), 310–317. https://doi.org/10.1177/08404704221108466

Wong, E. K. C., Thorne, T., Estabrooks, C., & Straus, S. E. (2021). Recommendations from long-term care reports, commissions, and inquiries in Canada. *F1000Research, 10*, 87. https://doi.org/10.12688/f1000research.43282.2

Woodhead, E. L., Northrop, L., & Edelstein, B. (2016). Stress, social support, and burnout among long-term care nursing staff. *Journal of Applied Gerontology: The Official Journal of the Southern Gerontological Society, 35*(1), 84–105. https://doi.org/10.1177/0733464814542465

Zhang, F., & Kaufman, D. (2016). A review of intergenerational play for facilitating interactions and learning. *Gerontechnology, 14*(3), 127–138. https://doi.org/10.4017/gt.2016.14.3.013.00

Zimmerman, D. R., Karon, S. L., Arling, G., Clark, B. R., Collins, T., Ross, R., & Sainfort, F. (1995). Development and testing of nursing home quality indicators. *Health Care Financing Review, 16*(4), 107–127. PMID: 10151883.

Zimmerman, S., Bowers, B. J., Cohen, L. W., Grabowski, D. C., Horn, S. D., Kemper, P., & THRIVE Research Collaborative (2016). New evidence on the Green House Model of nursing home care: Synthesis of findings and implications for policy, practice, and research. *Health Services Research, 51*(51), 475–496. https://doi.org/10.1111/1475-6773.1

Zimmerman, S., Dumond-Stryker, C., Tandan, M., Preisser, J. S., Wretman, C. J., Howell, A., & Ryan, S. (2021). Nontraditional small house nursing homes have fewer COVID-19 cases and deaths. *JAMDA, 22*(3), 489–493. PMID: 33516670. https://doi.org/10.1016/j.jamda.2

Rehabilitation 6

The Challenge

Older adults represent a wide range of health states. Maintaining one's health to experience a high quality-of-life is in everyone's best interest. If an older adult's health is compromised having access to and participating in a rehabilitation program to restore one's health is essential. The challenges that older adults face range from access to rehabilitation services, financial constraints, social isolation, physical and cognitive limitations and communication barriers (Fig. 6.1).

Long wait times for access to rehabilitation because of the limited availability of services coupled with geographic barriers can exacerbate the ability of older adults to seek supports that might prevent conditions worsening. The negative consequences of not getting the right support, at the right time from the right professional can negatively impact their quality-of-life. An older adult recovering from an illness, injury, loss of a spouse is at risk of experiencing social isolation. If they are lacking a support system that older adult may be challenged to engage in rehabilitation services and comply with a care plan. This can result in a poor prognosis for recovery and outcomes.

In many countries access to rehabilitation services that are medically necessary may be provided as a universally accessible program however if the previous two barriers to access are experienced it may be necessary to seek services outside of the public system. Financial constraints can be a limitation because the individual may not be able to pay for the services, costs of transportation or other expenses such as supplies or equipment that may be necessary for effective rehabilitation.

Older adults may also experience mobility issues, chronic pain or other health challenges that impact their ability to access community services. In addition to physical limitations the individual may also experience cognitive challenges that limits their ability to participate in what would be restorative activities. Aging adults experiencing hearing,

D. Juzwishin et al., *AgeTech Innovations in Healthcare for Older Adults*, Synthesis Lectures on Technology and Health, https://doi.org/10.1007/978-3-031-86885-6_6

Fig. 6.1 Woman exercising, supervised by doctor. *Credit* Shutterstock: DC Studi. https://www.shutterstock.com/image-photo/portrait-doctor-holding-tablet-while-elderly-2450842353

vision, language barriers or experiencing cognitive decline may also face barriers in being able to communicate with health care providers, comply with their prescribed treatment plans and engage in restorative activities.

Persona and Scenario—Jin

Jin, an 86-year-old woman, lives alone and has just been discharged from a longer-than-expected hospital stay, after the investigation of her cardiac arrythmia led to a hospital-acquired Clostridium difficile infection. Jin experienced symptoms consistent with delirium while she was admitted, and her cognitive abilities are declining. Home care and rehabilitation therapy helped her to regain some mobility in the month after she returned home, but one day Jin had a sudden stroke while she was greeting her daughter Asha at the front door. Asha heard her mother slur her words, saw her slump onto the entryway bench and Asha phoned for an ambulance, fearful of her mother's recent arrythmia. At the hospital, Asha learned that a clot had stopped blood flow to her mother's brain and that the damage had been severe. After four weeks in hospital, Jin is awake and can sit up on her own. Her speech is not easily understood, and she still eats liquids as she experiences right-side facial droop and paralysis (hemiplegia) on the right side. Given the severity of her symptoms, and her relatively good health prior to her recent hospitalizations, Jin is assessed as having high restorative potential and will receive in-patient rehabilitation at the hospital (Rehab Care Alliance, 2014). Since this service occurs in the hospital, she will not pay for these services. After six weeks of physical rehabilitation, Jin has regained some mobility in her right arm, and her facial droop has lessened enough so she can eat solid food. Jin's speech is hard to understand. She is being discharged home and is eligible for one day of rehabilitation service through physiotherapist home visits to continue her therapy. Asha tells her mother that she read online that the more rehabilitation she completes in the period immediately after the stroke, the better chance she will have to continue to restore her function to her pre-stroke levels. If she were to seek out more service, she would have to pay for it, and she is not sure how she would get to appointments, as she does not drive and is not comfortable taking public transport, especially now that her speech is hard to understand, and is more easily confused by new things.

How Can Rehabilitation Help Older Adults?

Older people, like Jin, require the functional ability to complete activities of daily living to fulfill their overwhelming desire to age-in-place in their own homes (Means, 2007; Peek et al., 2016). Higher rates of chronic disease are associated with greater risk of disability, and the greater one's inability to independently complete the tasks required to age-in-place at home, the more likely that person is to require institutional care (Oh et al., 2019; Sheridan et al., 2019). The increased prevalence of chronic disease in older people threatens the goal of aging-in-place by impairing functional ability to complete activities of daily living (CIHI, 2011; Marengoni et al., 2011).

There is good evidence to suggest that multidisciplinary rehabilitation services for older people reduce frailty, hospital readmission and placement to LTC (Cameron et al., 2013; Cameron et al., 2021; Dyer et al., 2020). We define multidisciplinary rehabilitation as multiple providers working as a team, that in their care, may overlap with activities or patients normally dealt with by other disciplines, but approach rehabilitation activities from within their discipline (Melvin, 1980).

Rehabilitation services including physiotherapy, orthotic devices, water therapy and integrated rehab programs have been shown to reduce further health care costs and improve clinical outcomes, prevent surgeries and increase labor productivity (Shahabi et al., 2021). A recent review of economic analyses of rehab services for osteoarthritis patients showed that "from all perspectives, physical rehabilitation interventions were valuable economically" (Shahabi et al., 2021, p. 193). As people living with osteoarthritis are more likely to be older, and have multiple co-morbid conditions (Swain et al., 2020), the findings from the Shahabi et al. (2021) study provide relevant rationale for the effectiveness of providing rehabilitation to older people with complex health issues.

Assistive technologies are an important tool to facilitate the process of 'making fit again' and are defined by the WHO as devices designed to "maintain or improve an individual's functioning and independence to facilitate participation and to enhance overall well-being" (2021a, para.1). Technology can help older people managing multiple chronic conditions and preserve functional abilities that are the foundation for independence and aging-in-place (Hajat & Stein, 2018; Peek et al., 2016). This chapter will outline the role for AgeTech to support the care and quality-of-life of older people receiving rehabilitation services. The chapter focuses on three categories of AgeTech used in rehabilitation: robotics; virtual reality; and smartphone-based apps to manage chronic disease.

The WHO estimates that 2.4 billion people worldwide have health conditions that could benefit from rehabilitation, and yet: 'the need for rehabilitation is largely unmet' (2021b. para. 16). Unmet need for rehabilitation services is exacerbated in low- and middle-income countries where more than half of people do not receive the rehabilitation services they need (WHOb, 2021). Based on the Global Burden of Disease, Injuries and Risk Factors Study, a comprehensive assessment of prevalence, incidence and years of life lived with a disability for > 300 diseases and injuries for all countries, the need for rehabilitation services is estimated to have increased by 63% between 1990 and 2019 (Cieza et al., 2020).

Unmet need for assistive technologies (a broad term which includes AgeTech) to help manage disabilities or chronic diseases is equally vast. In 2010, the *World report on disability* projected that by 2050, more than two billion people would require an assistive technology (WHO, 2011). Living without access to devices leaves disabilities uncompensated for, isolating people from normal daily activities, work and leisure, with a stark impact on quality-of-life (Alqahtani et al., 2021).

Another factor contributing to broad demand for rehabilitation services is that recovery depends on intense and repeated task specific practice (Backus et al., 2010, p. 362), which is time and resource intensive to provide. The practice must be challenging and engaging to demonstrate effects in patients (Bishop et al., 2020). Other factors such as weather, transportation and long distance to travel complicate patients' abilities to participate in outpatient rehabilitation service (Sheehy et al., 2019). In many places, rehabilitation staff are in short supply, services are wait listed and each patient is offered limited appointments (Krishnaswami et al., 2020; Sheehy et al., 2019). In Canada, there are 7.02 practising physiotherapists per 10,000 people; which is half as many as in Australia (14/10,000); and more than three times fewer than in Germany (24/10,000) (World Physiotherapy, 2022). The United States and the United Kingdom have 9.56 and 8.8 practising physiotherapists per 10,000 population (World Physiotherapy, 2022). The shortage of physiotherapists is exemplified by the fact that Canadian physiotherapists have an unemployment rate < 1% (Conference Board of Canada, 2017). The short supply of physiotherapists makes care challenging. For example, in the United Kingdom, a survey covering 71% of in-patient stroke units, demonstrated that most of the units are operating below Department of Health guidelines, limiting their ability to provide the recommended amount of therapy for optimal functional recovery for stroke patients (McHugh & Swain, 2014).

Spa Therapy and Health Resort Medicine

Health resort medicine, which can include balneotherapy, hydrotherapy, climatotherapy, and can be mixed with exercise, relaxation therapy and other physical therapy is common in several European and Asian countries (Stier-Jarmer et al., 2015). Germany's comparatively high rate of physiotherapists may be related in some part to their long tradition of health resort spas, many of which are dedicated to rehabilitation. Until around 1990, some of these services were covered through Germany's national health insurance program—this meant that health care providers such as doctors or physiotherapists working at these spas could prescribe spa resort treatment to patients at cost to the German health care system (Pillmayer et al., 2021). German governments have since reformed health insurance coverage, and now spa resort treatments are part of a 'second health market,' in which the payers of these services are individual patients or workplace health insurance plans (Pillmayer et al., 2021).

The health and wellness spa industry attracts more than 100 million visitors annually, and is an industry valued at more than 1 trillion in American dollars (Erfurt-Cooper & Cooper, 2009). Total spa revenues in the United States in 2021 were $18 billion dollars, only 5% below pre-pandemic revenues (International Spa Association Foundation, 2022). There is growing interest in leveraging ICT technology and artificial intelligence (AI), within the health resort medicine field, e.g., digital booking, growing ecommerce, providing self-guided or technology driven spa-experiences (Erfurt-Cooper & Cooper, 2009; International Spa Association Foundation, 2022). The 'baby boom' generation has been cited as driving growth in the health resort medicine field (Erfurt-Cooper & Cooper, 2009). Growing trends of individuals being attune to and willing to pay for health prevention services, combined with a growing evidence base of health resort medicine seem to bode well for the smart technology enabled health resorts of the future (Erfurt-Cooper & Cooper, 2009; Stier-Jarmer et al., 2015).

AgeTech and Rehabilitation

The word 'rehabilitation' originates in Latin and combines the terms 'to make fit' and 'again.' The WHO defines rehabilitation as *"a set of interventions designed to optimize functioning and reduce disability in individuals with health conditions in interaction with their environment" (WHO, 2021). Other definitions have focused on the role rehabilitation plays in maintaining physical function in older people (*Crocker et al., 2013*). At the base of* *rehabilitation is the commitment to retaining as much of people's independence as possible in the everyday activities that matter to them: learning, working, caring and recreation (WHO, 2021). These activities are commonly referred to as 'activities of daily living' (ADLs) that include eating, drinking and bathing* (*Hopkins* et al., 2017). *Instrumental activities of daily living are a more complex type of task which underpin the ADLs, for example, managing money to buy food, transporting it and cooking,* etc. (Hopkins et al., 2017).

Rehabilitation can be active (specific exercises or physical activity) or passive (preformed on a client), it can be group-based or done individually, and it can be led by a rehabilitation professional or self-directed (Crocker et al., 2013). *Rehabilitation normally unfolds in four steps assessment to determine the extent of the patient's limitations; goal-setting to guide treatment and planning the intervention; implementing the intervention by practicing movements over time; and re-assessment to determine new goals* (Laut et al., 2016).

Rehabilitation Services in Canada

Who provides rehabilitation? Rehabilitation services are commonly delivered by audiologists, chiropodists, occupational therapists, physiotherapists and speech-language pathologists and may involve other health care providers as part of a team-based approach (nurses, geriatric and physiatry specialists, pharmacists, dietary and or psychiatric professionals and other assistants) (MOHLTC, 2014)Different jurisdictions may have different norms by which health professionals engage in rehabilitation provision, and whether they are regulated health professionals who are licensed by a college (e.g., Health and Care Professions Council, the United Kingdom's physiotherapist regulatory body) or not (e.g., personal support workers). Though many different types of health care workers can provide rehabilitation services, they share an orientation towards collaborative care that meets the needs of patients and their families (WHO, 2017). Core rehabilitation competencies include compassion, respect for diversity, dignity and self-determination towards the patients served (WHO, 2017).

Who needs rehabilitation? In Canada, in 2017–2018, the most common reasons for requiring inpatient rehabilitation were: orthopedic conditions and stroke, and older adults aged 65 + represented 75% of Canadian cases (CIHI, 2021). Canadian females make up a larger share of inpatient rehabilitation clients (53%) than men, potentially reflecting their longer life expectancy (84 years) compared to men (80 years) (The World Bank, 2019).

Where are rehabilitation services provided? Physical rehabilitation services are provided across the continuum of care including in specialized rehabilitation hospitals, acute care hospitals, in individual homes (facilitated by home care workers), in LTC homes, and by rehabilitation staff in private clinics. Most rehabilitation services occur outside of hospitals in Canada (Sheehy et al., 2019).

Geriatric rehabilitation is multidisciplinary set of interventions designed to restore functional ability and enhance participation of older people following a period of acute decline (Lohr., 1997; Romke van balen et al., 2019). Geriatric rehabilitation commonly occurs after a cognitive geriatric assessment, which is a 'multidimensional process to identify medical, social and functional needs and the development of an integrated care plan to meet those needs, involving medical/physical, psychological/psychiatry, socio-economic, functional and nutritional assessment' (Parker & Geron, 2007, p. 150). Geriatric assessment helps to identify older people's needs, and geriatric rehabilitation is the multidisciplinary provision of services in order to meet older people's needs in a holistic way that consider more than simply their physical ability.

AgeTech has the potential to support older adults' daily activities, to support aging-in-place and increased independence for a growing older adult population (Agree, 2014; Lansley et al., 2004; Wilson et al., 2009). Rehabilitation services enable older people to manage chronic conditions and maintain functional abilities required to age in a place of their choice, a goal that is shared by governments as it decreases the need for institutional support, traditionally provided by the state. AgeTech has been used in palliative care; chronic disease management; mental health and behavioral health; supports for family caregivers; and transitional care for heart failure and home-based primary care for frail older adults (Quinn et al., 2018). AgeTech has demonstrated impacts on older adults' quality-of-life: reducing pain, increasing cognition, and increasing the likelihood of patients completing rehabilitation activities independently (Papi et al., 2016; Tomšič et al., 2018). AgeTech has shown impact on important health system indicators as well reducing health service utilization (Emergency room and rehabilitation admission) and shortening the length of hospital stays (Martin-Lesende et al., 2013). To support older adults' wishes to age in a place of their choosing, their functional ability to complete activities of daily living must be preserved and enhanced. Rehabilitation services delivered or facilitated by AgeTech can support older people's functional ability and thus warrant further investigation.

Two prominent categories of AgeTech used in rehabilitation that are discussed in this chapter are robotics and virtual reality. Challenges and opportunities of each type will be discussed, but this should not be considered an exhaustive list of AgeTech for rehabilitation. Each of these types of technology help to surmount identified challenges associated with traditional rehabilitation. These challenges include its time consuming nature as well the its labor-resource intensiveness and its dependence on patient compliance. As it has limited availability in certain geographies, success is reliant on transport to centers, coverage beyond initial phase of treatment is costly for patients (Saposnik et al., 2011).

Robotics and Rehabilitation

This section will cover some of the available research evidence on those robotic systems for use to maintain, restore or improve physical function. The International Federation of Robotics (IFR), a global association of national robot industry groups, distinguishes between industrial and service robots (Dixon, 2020; IFR, 2021). Industrial robots are mostly computer-controlled arms with attachments that serve to manipulate the physical world, and have been dubbed "AI with opposable thumbs" (Dixon, 2020). Service robots can navigate the world independently and help people to complete tasks such as diagnostics and rehabilitation (Dixon, 2020; IFR, 2021). Robotics for rehabilitation, including wearable robotics such as exoskeletons, can help provide the repetition and intensity of treatment that is required to see results (Banquied et al., 2021). The prevalence of exoskeleton research is significant- a recent review found that exoskeletons made up 33% of artificial intelligence research for older adults (Ma et al., 2023). Amid a shortage in rehabilitation staff and increasing demand for rehabilitation services among a growing population of older people, robotics may be a promising tool to assist staff in rehabilitation activities for older people. This section will exclude the wealth of information on a third type of robot-humanoid, targeted towards the social isolation of older people (Góngora Alonso et al., 2019).

Key Initiative. I CAN Centre for Assistive Technology Services, Glenrose Rehabilitation Hospital, Edmonton, Alberta, Canada
Older adults in four Canadian provinces/territories can be prescribed access to AgeTech located in the I CAN Centre, part Canada's largest rehabilitation hospital, the Glenrose. At this center, patients, families and care partners receive access to state of the art technology and a multidisciplinary team of rehabilitation experts including occupational therapists, physical therapists, speech-language pathologists, rehabilitation engineers and/or assistive technology specialists. The care team meets with the patient to understand their goals for rehabilitation, and together they write a plan that includes solutions that can be no-tech, low-tech or high-tech. The care plan includes access to a device loaning library and technical support (AHS, n.d.). Access to this program is provided through referral from a health care provider.

The cost of robots for rehabilitation cannot be overlooked. Earlier studies have shown exoskeletons to cost in the tens of thousands of dollars (Robinson et al., 2013; Stefanov et al., 2004). Cost continues to be extremely variable, as demonstrated by a simple Google search for a wrist robot, which shows the same photos of a "rehabilitation robot glove" priced at $75 or $780. 05, at the time of this chapter's writing, depending on the retailer (Ali Express, n.d.; DHgate, n.d.). Older people and/or their families looking to purchase robotic systems may have significant challenges negotiating effectiveness and value for money with robotic systems operating outside clinically prescribed environments. For example, an exoskeleton for lower-limb rehabilitation needs enough power and dexterity to support patients to train muscles and move with a speed and force similar to the real-world. Robotics for lower-limb rehabilitation require the power to thrust nearly 250 pounds (1000 N) to account for carrying the potential load of patient of the same weight (Yurkewich, 2015). Meeting these demands requires exoskeletons to use motors that are powerful, heavy and costly, to support patients who can be heavy, strong and/or rigid (e.g., patients after stroke), which contributes to their high costs (Yurkewich, 2015). Strategies to reduce device complexity have been suggested to lower costs of robotics (e.g., low-cost sensors, light-weight motors) (Yurkewich, 2015). Robotics that provide rehabilitation to one joint at a time, potentially through seated robotics, may be less costly than robotic gait training, which is the simultaneous training of multiple joints and muscles movement required for walking (Saglia et al., 2012; Yurkewich, 2015). McKinsey & Company's *Industrial robotics: Insights into the sector's future growth dynamics* report shows that reducing costs for robotics is a significant priority industry wide (Teulieres et al., 2019). Rehabilitation practice for older people stands to benefit from reductions in costs of this technology.

Case Box. ReWalk Exoskeleton

CADTH, Canada's agency for Drugs in Technology and Health, is mandated is to provide objective evidence to support health system decision-making. CADTH featured the ReWalk powered gait orthosis for lower limb impairment (manufactured in the United States) in their series "Issues in emerging health technologies" in 2015. This device is the first exoskeleton to be approved by health regulators for personal use (Murtagh, 2015). Their summary found the tech was safe and could help people with mobility problems to sit, stand, walk, turn and use stairs. This profile noted the high cost of the device $85,000, in American funds, plus additional maintenance fees for an institutional license, with a lifespan of five years (Murtagh, 2015). American and Canadian health technology regulators reclassified powered exoskeletons as higher risk between 2013 and 2014, meaning they were subject to higher regulatory control/standards. In the United States where ReWalk is approved for personal use (as opposed to institutional use), its use requires the assistance of a trained companion. For this reason, there is concern about the use of ReWalk increasing the need for health care aides, as it cannot be used independently. In CADTH's review, six observational studies, funded by the manufacturer of the device, showed that use of this device did not allow participants (spinal cord injury patients) to reach a gait speed sufficient for community use of ReWalk (Murtagh, 2015). On average, patients trained on the use of this device for 24 60–90-min sessions, and while they were able to walk, they were not able to manage changing walking surfaces, heavy foot traffic, or to walk quickly enough to cross a street

(Benson, 2016; Murtagh, 2015). None of the study participants were older people. There are two ReWalk systems being used in rehabilitation centers in Canada, and more data is required to know if/how these devices can function outside of rehabilitation institutions (Murtagh, 2015).

There is research evidence to suggest AgeTech robotics are effective for use in the rehabilitation of older people. For example, a balance assistive robot led to significant positive changes in three balance measures among older people with hip fractures (Takano et al., 2020) In some cases, where high intensity of therapy was required (stroke rehabilitation), robots preformed better than physiotherapists (Crea et al., 2017). Another study has shown that in 80% of included studies, robotic-assisted rehabilitation was found to be more cost-effective than the same dose of conventional therapy for stroke patients (Lo et al., 2019). The effects in this study were more pronounced among patients who had severe stroke-related disability i.e., those who required high intensity of therapy (Lo et al., 2019). Similarly, in other neurological disease studies (such as Parkinson's disease, stroke and spinal cord injury) robot assisted gait therapy showed improvements in gait and balance, compared with traditional rehabilitation (Furnari et al., 2017; Picelli et al., 2012; Sale et al., 2012). Alwardat et al. (2018) completed a systematic review and meta-analysis with pooled results from seven randomized-controlled trials demonstrating that robot-assisted gait therapy showed significant improvements in balance, walking distance, stride length and a Parkinson's disease rating scale. In 2021, a systematic review and meta-analysis of 20 randomized controlled trials of nearly 800 patients from more than 10 countries, found similar improvements in walking ability, endurance, and speed among older stroke patients who completed robotic exoskeleton training. Balance, stride length, speed and endurance are essential skills that form the basis of functional ability to live independently.

The main challenges associated with robotics for rehabilitation are availability and cost. Robots have been criticized for their slow movement through the development process, with some authors noting that AgeTech do not (or does not) respond to real life problems of older people when they get out of the lab (Bedaf et al., 2015). Over 20 years ago, Cooper and Keating pointed out that there is no 'rehab store' where older people can go to view, test and try technology for rehabilitation (Cooper & Keating, 1996). There has been limited progress on this front, with busy rehabilitation staff the primary gatekeeper to AgeTech for rehabilitation, especially in Canada. The opposite is true in Norway, where 'assistive technology centres' exist for patients to visit to purchase a variety of assistive technology products, learn from trained staff and receive repairs or refurbished technologies which are all certified to meet the country's quality standards (Sund, 2017; WHO and UNICEF, 2022). In the province of Ontario in Canada, there are government programs to reimburse older people for purchases of assistive technologies such as eyeglasses, walkers or wheelchairs, but no robots for use in rehabilitation appear on this list (Ministry of

Health, 2022). Australia and New Zealand are other high-income countries with an even more restrictive list of government reimbursable assistive devices (Marchildon & Peckham, 2019). Robots for rehabilitation spend long time periods in laboratory-like settings before becoming commercially available and when they do become marketed for purchase, there is not a central access point to acquire robotics for rehabilitation or financial support for this category of technology. The average older person in Canada relies on a busy rehabilitation staff to prescribe them access to a robotic system, usually located in a rehabilitation clinic or hospital.

In other studies where robot assisted therapy was not more effective than usual therapy, it showed promising results in other meaningful outcomes. For example, two randomized controlled trials with stroke patients did not show robotics as better than usual therapy (Alingh et al., 2021; de Luca et al., 2020). Though, in one of these studies, robot assisted gait therapy showed a marked difference in certain gait abnormalities for the patients who received robotic training (Alingh et al, 2021). This is a significant finding as gait abnormalities are one of the most common rehabilitation goals (Alingh et al., 2021). In another trial where robotic assisted upper limb therapy was not superior to usual therapy for stroke patients, the robotic therapy showed better recovery scores among the oldest age groups in their trial (71–80 and 81 + years old) (Cecchi et al., 2021). Among those receiving usual therapy, as age increased, recovery scores decreased (Cecchi et al., 2021). In another trial where an overground exoskeleton was used in rehabilitation with stroke patients, it showed positive impacts on constipation, mood and coping strategies (the group receiving usual therapy did not experience these changes) (de Luca et al., 2020). Taken together, these results show that even in well-designed studies where robotic assisted rehabilitation did not prove to be superior to traditional rehabilitation, there can be important results on outcomes important to older people gait abnormalities; rate of recovery from stroke; and health related quality-of-life.

Virtual Reality

Virtual reality (VR) has been described as is a way for people to visualize, manipulate and interact with computers and complex data, or as an advanced form of human-computer interaction, allowing users to interact with and immerse themselves in a 3D virtual environment (Aukstakalnis & Blatner, 1992; Schultheis & Rizzo, 2001). In rehabilitation, these virtual environments can help to assess and restore cognition and function by simulating the tasks of daily living. VR in rehabilitation can be enabled by commercial systems like Nintendo's Wii, Xbox's Kinect or there are devices designed especially for use in rehabilitation (Dockx et al., 2016). Some authors disagree with the conceptualization of the Wii or the Kinect as VR when they operate in two dimensions calling them "desktop VR" (Tieri et al., 2018). According to these authors, only technology that operates in three dimensions should be labeled as VR (Tieri et al., 2018). Given there are many usability, feasibility and systematic reviews of such studies which classify 2D devices as VR and this chapter will consider them as VR.

VR has been used successfully in a variety of areas of health care including in pain management to distract patients from wound rehabilitation, chemotherapy or dental work (Schultheis & Rizzo, 2001). Other areas of health care where VR has been successful are in medical education, surgical procedures and with mental health patients to treat body image issues and post-traumatic stress disorder (Schultheis & Rizzo, 2001). There has been anticipation about the role for VR in assessment and measurement in rehabilitation and the role it can play to standardize therapy, track progress and ensure that the 'dose' of therapy is just challenging enough to demonstrate results (Laut et al., 2016; Schultheis & Rizzo, 2001). VR can promote safer task practice than real-life allowing for transfer of mental and physical skills to 'real-life' (Fung & Keshner, 2017). The most promising aspect of VR in rehabilitation is the motivation it can provide patients to complete repetitive exercise to achieve maximum restoration of functional ability (Dockx et al., 2016). The immersive nature of VR effectively distracts users from the fact that they are exercising. One systematic review confirmed that VR used for rehabilitation led to better adherence to exercise and compliance with rehabilitation programs (Wang et al., 2019). Especially in the case of stroke, traditional rehabilitation has been described as boring, and falls short of desired impacts (Shi et al., 2018). In one VR study, older adults found that exercise was more enjoyable as they were playing games with family and friends (Kerkhof et al., 2016). The social component of rehabilitation technology BikeAround is a key feature of the technology, the idea is that older people with dementia, might choose to cycle through an area familiar to them while conversing with a therapist or caregiver about what they see, triggering memories (Martel et al., 2018).

Virtual reality has been used successfully to overcome one of the main issues in all rehabilitation—adherence to rehabilitation activities. Like the availability issue affecting robotics and exoskeletons, VR has been said to be more common in research than practice (Dockx et al., 2016). Where available, these technologies are available for commercial use (as opposed to individual use) and can be found at a few rehabilitation centers or LTC homes in Canada (Martel et al., 2018; Sheehy et al., 2019). As VR AgeTech has not yet been consistently used in rehabilitation practice, there are no protocols that define the volume, intensity, time, or specific tasks to be used for different types of patients (Porras et al., 2018). Clear protocols around the principles of VR use for rehabilitation of older people would help therapists integrate this technology in their everyday practice.

A drawback to the use of virtual reality in rehabilitation is that it still involves therapist assistance. VR necessitates therapists learn how to use VR and integrate it in their practice, know how to convey these instructions to their clients, and in many cases, follow along as their patients use VR, providing motivation and correcting movement patterns that might be problematic (Porras et al., 2018). Similar to ReWalk exoskeleton, some VR for rehabilitation such as BikeAround or JinTronix require a rehabilitation therapist or trained family caregivers to help older people use these technologies (Martel et al, 2018;

Sheehy et al., 2019). VR is an important tool for therapists that can enhance older people's adherence to rehabilitation protocols but it does not replace therapists, or free their time to work with more clients.

BOX 1: JinTronix virtual rehabilitation technology

Virtual Rehabilitation Technology JinTronix virtual rehabilitation technology uses Xbox Kinect to offer evidence-based movements, virtual games with continuous live video and audio feedback and can be used in personal or institutional settings (Martel et al., 2018). By correcting their movements in response to feedback, users can progress to higher levels. Speed, range of motion, number of repetitions and duration can be tailored to suit different users and JinTronix tracks the quantity and quality of movement through an online portal that kinesiologists or therapists can access (Martel et al., 2018). In one study comparing an exercise program delivered by JinTronix with the same program delivered in-person, the JinTronix group showed significantly greater walking speed improvements, and an adherence rate nearly 10% higher than the in-person group (Martel et al., 2018). At the time of publication, Jintronix had rebranded as Ludica Health, releasing LudoFit and LudoFit PRO, a subscription-based app, which uses the camera of a computer or tablet to track the user's movements which are simulated by a character on the screen (Ludica Health, 2024). A user doing weight shifts and squats is played back as their character skiing down a mountain in Italy. LudoFit is for at-home use by individuals, and LudoFit PRO is a therapy exercise program for use by physical and occupational therapists treating clients (Ludica Health, n.d).

BOX 2: JDome BikeAround

Immersive Technology

JDome BikeAround is an immersive technology that uses google street view to cycle any address. In a project where long term care residents used BikeAround as a supplementary fall-prevention strategy in addition to a group-based fall prevention program, older people reported high rates of enjoyment using BikeAround, that were not reported for the group-based intervention (CABHI, 2022). The people who completed six weeks of BikeAround sessions and six weeks of group-based falls prevention activity showed a reduction in falls, and improvements in physical endurance and quality-of-life (CABHI, 2022).

How Could These Technologies Help Jin?

Some studies suggest that up to 20% of atrial fibrillation patients go on to have a stroke and up to 80% of stroke survivors suffer from upper-limb impairment after having a stroke (Alshehri, 2019; Hayward et al., 2019). Stroke patients are motivated in their rehabilitation to regain function, but the best dosage, timing and stroke rehabilitation activities are still unclear (Hayward et al., 2019). As Asha noted, evidence from animal studies shows that there is value in beginning rehabilitation for upper-limb impairment as early as ten days post-stroke when the neuroplasticity in the brain is high, which can facilitate better rehabilitation outcomes (Hayward et al., 2019). Jin is being provided pragmatic rehabilitation; the most the health system can offer him, as opposed to the amount that might work best to restore his function.

An upper-arm rehabilitation system that can offer a high intensity of reaching therapy, that adapts to his changing capacity, and even tracks his progress would be very helpful. Jin is not likely capable of purchasing technology, or supplementing his government-allocated therapy by paying out-of-pocket for more service. This scenario is likely common. If Jin could be prescribed access to the ReTouch system, she could gamify his rehabilitation experience which might make him more likely to complete it. A clinic with a robot-assisted upper-arm rehabilitation he could access would also help Jin, but even if he got access to these technologies, there are practical concerns for him, such as transportation to the clinic, and his self-consciousness around the stroke's after affects on his speech. Access to a JinTronix system that could operate in his home, with rehabilitation exer-games for her and tracked information for his therapist to monitor her progress could be very useful to Jin's post-stroke rehabilitation, even helping him to meet the recommended 15 h + /week of rehabilitation exercise required to increase function after stroke (Kwakkel et al., 2004). In 2023, a Jintronix system for commercial use is sold for $13,905 USD, but as of 2014, American patients can access Jintronix software for a $50 monthly subscription fee in addition to the cost of purchasing an X-box Kinect ($250) (Bass, 2014; Direct Supply Inc., 2023).

Lessons from the COVID-19 Pandemic

Although projections of need for rehabilitation services had often been assumed in Canada, based on its aging population, increasing rates of chronic conditions, changing discharge patterns, that need has never been quantified (Landry et al., 2007). This demand may be exacerbated by the COVID-19 pandemic, which demanded reallocation of financial and staff resources to manage surges of cases (Atashzar et al., 2021). Many rehabilitation services were transitioned to telemedicine, leaving rehabilitation's primary client base (older adults) lacking the support they needed (Atashzar et al., 2021; Leocani et al., 2020; Venketasubramanian, 2020). The need for rehabilitation services may be increased among older adults who have had COVID-19. One Canadian cohort study reported worsened mobility (in three domains) among community-dwelling older adults with a COVID-19 diagnosis, even though none of these patients required hospitalization (Beauchamp et al., 2022). COVID-19 has exacerbated need for rehabilitation among those already receiving it and created another disease category that may necessitate rehabilitation services.

Telerehabilitation is defined as a tele-medicine subfield consisting of a system to control rehabilitation at a distance (Zampolini et al., 2008). Since the impact of the COVID-19 pandemic isolated people from access to face-to-face services there has been interest in better understanding if telerehabilitation is an effective alternative. In a Cochrane systematic review, Laver et al., (2020, p. 3) state:

> We found that people who received telerehabilitation had similar outcomes for activities of daily living function to those that received face-to-face therapy and those that received no therapy (usual care). At this point, not enough research has been done to show whether telerehabilitation is a more effective way to provide rehabilitation. Some studies report that telerehabilitation is less expensive to provide but information is lacking about cost-effectiveness.

> Only two trials reported on whether or not any adverse events had occurred; these trials found no serious adverse events were related to telerehabilitation. Further trials are required.

This finding opens the door for imagination and creativity to identify how AgeTech coupled with program developments in telerehabilitation could significantly improve the outcomes for older adults requiring rehabilitation, but are restricted in access.

Engaging Patients in AgeTech for Rehabilitation

Older people receiving rehabilitation services may present with complex health challenges, which, as noted in earlier chapters, may be exacerbated by health care that is organized and delivered in disease-based 'silos.' Complex health challenges, combined with a complex system offering rehabilitation in different care settings (acute care, LTC and homecare) with different levels of coverage, depending on age and geography (Mattison et al., 2020) create a challenging environment in which to integrate AgeTech. Technologies developed with input from end-users, be it older people, caregivers, rehabilitation therapists and/or health system decision-makers can help to ensure AgeTech fit within complex health systems (MacNeil et al., 2019). Despite the crucial importance of patient perspectives (Nasr et al., 2016; Robinson et al., 2013; Wang et al., 2017) and increased advocacy for inclusion of patient engagement in technology development, promising health technology innovators may not understand how to meaningfully include patient voices in technology development (CADTH, 2022). The Reverse Trade Show may be a useful forum to convene patients, clinicians, decision-makers and innovators and build relationships to inform the development of patient-centered AgeTech. One of the authors (Axel Meisen), is involved with the development of rehabilitation technology at the Glenrose Rehabilitation Hospital in Canada for several years. Experience is that although good technologies were conceptualized and developed none have been commercialized on a large scale. The challenge is that companies that produced the technology were too small and lacked the expertise to create and serve large markets.

Opportunities: AgeTech Training for Rehabilitation Patients and Providers

Access to AgeTech for rehabilitation can be limited by the need for a referral by a rehabilitation professional, in this way, rehabilitation providers are like gatekeepers of AgeTech in rehabilitation. At the interface of patients, their families and their physical function, rehabilitation providers are in an excellent position to recommend and refer older people to AgeTech, but they face numerous challenges in doing so. It can be challenging for providers to learn which AgeTech are effective for which patients, not to mention the burden of staying abreast of new technologies that could have implications for their patients (Myburg et al., 2017). A train-the-trainer approach to continued education for rehabilitation providers on clinically effective AgeTech to aid practice, has been suggested as a way to leverage the important role of rehabilitation providers in mediating access to AgeTech (Myburg et al., 2017). Continued education certification on AgeTech could be focused on AgeTech that is reimbursable by governments or community groups.

For AgeTech that is not reimbursable by governments, there may be a role for governments to act as an information broker about AgeTech. This would provide a central location for unbiased review and rating of AgeTech for rehabilitation that are not reimbursable, to protect the public from purchasing costly rehabilitation tech for which there may not be good evidence of effectiveness. We know from the literature that families in crisis turn to independently purchasing technologies to help them care for family members (Gélinas-Bronsard et al., 2019), governments can provide an oversight function to ensure families in crisis do not purchase ineffective AgeTech. As an example, AGE-WELL worked with academic researchers and Alzheimer's Societies in Canada to create evidence-based information about locating devices for people with dementia with a history of wandering or becoming lost. This included criteria for choosing devices, and a list of technologies for purchase (Alzheimer Society of Canada, 2022). A similar website could be created to streamline effectiveness and purchasing information about AgeTech for rehabilitation in cases where those technologies are not reimbursable by governments.

AgeTech Training for Older People and Their Families

As seen with exoskeletons and virtual reality technologies, training of family and friend caregivers and/or rehabilitation assistants is required to ensure the safe use of AgeTech for rehabilitation. Older people themselves may require training to ensure the safe and continued use of AgeTech to support rehabilitation goals. Training older people about the use of AgeTech for rehabilitation should be tailored based on the health status of the older person (Kerkhof et al., 2016) and pre-discharge from hospital may be an important time to offer older people training and practice with AgeTech (Myburg et al., 2017). Role-playing and take-home resources containing troubleshooting information about AgeTech can supplement hands-on practice to increase the use of AgeTech for rehabilitation outside institutional settings (Darcy et al., 2017; Reyes et al., 2018; Seaborn et al., 2016).

Low-cost AgeTech for Rehabilitation

The incredible costs associated with promising AgeTech in rehabilitation, such as exoskeletons and virtual reality devices, cannot be overlooked. Although lower cost robotics are an identified priority area for the health technology industry (Teulieres et al., 2019), delivery of rehabilitation services via widely available technology, such as smartphones has been suggested as a method of mitigating high-cost AgeTech (Kouroupetroglou et al., 2017). Future research is required to demonstrate the cost avoidance to health care systems of AgeTech for rehabilitation. Given that more than a quarter of the world's adult population (1.4 billion adults) are insufficiently active and the fact that there has been no improvement in global levels of physical activity since 2001 (WHO, 2020), there is incredible potential for AgeTech in rehabilitation to gamify and motivate older people to be more active, which could translate to significant cost-avoidance to health care systems.

Seminal Demonstration Project

Key Initiative. Reverse Tradeshow, W21C Research and Innovation Centre, Calgary, Alberta

The W21C (Ward of the Twentyfirst Century) Research and Innovation Centre is a partnership between the University of Calgary and the Calgary Zone of Alberta Health Services in Calgary, Alberta. A Reverse Trade Show reverses the format of a traditional trade show, with health care providers from continuing care settings providing brief presentations on the challenges they face on a daily basis. After presentations, providers can meet innovators, entrepreneurs and industry attendees to discuss technological solutions to problems (O'Brien Institute for Public Health, 2022).

Reverse trade shows surmount the issue within AgeTech of technology being developed without input from the care settings, patients and care partners and subsequent troubles gaining traction within health systems and prospective purchasers (MacNeil et al., 2019). For example, a company called Rehabtronics discovered the ReTouch technology through a Reverse Trade Show and struck a deal to manufacture this technology globally (AHS, 2013). ReTouch was a technology developed in partnership between the University of Alberta and the Glenrose Rehabilitation Hospital, where students helped to modify an adjustable screen-table arm rehabilitation game that becomes more difficult as patients make progress (CBC Edmonton, 2013). Patients who require motor limb therapy use ReTouch to gain strength, improve coordination and reaction skills. News releases describe patients dripping in sweat, not realizing they had spent half an hour doing rehabilitation exercise. The technology system retails for $16,000 Canadian dollars and six are available for use at the Glenrose Rehabilitation Hospital (AHS, 2013).

A Virtual Reverse Trade Show took place in July 2020, with the key challenge being related to COVID-19, specifically, communicating with staff for contact tracing; enhancing staff and resident well-being, and education and reinforcement of infection prevention and control. Sixty- two health system stakeholders, clinicians, innovators and entrepreneurs took part in the trade show. Nine technological solutions are being modified to address challenge areas, after partnering at the Virtual Reverse Trade Show. The technologies being pursued include a light projection game for people with dementia, an audit service for infection prevention and control in LTC, a chatbot to link isolated older people to volunteer for conversation, and an app to allow families of people living in LTC to see care tasks being completed in real time (O'Brien Institute for Public Health, 2022).

Concluding Remarks

As Dr. David Naylor noted, and he is quoted in the Glenrose Rehabilitation Hospital Strategic Plan, "Much of modern health care is indeed rehabilitation in one form or another" (AHS, 2018). As scientific progress continues to lengthen lifespans of older people, the management of chronic conditions will increase in importance. AgeTech can play an important role in increasing motivation and adherence to activity to support rehabilitation goals and has been proven to be as effective as usual rehabilitative care across a range of outcomes and conditions that afflict older people. In some cases, AgeTech has shown to be more effective than usual rehabilitative care. These impressive results can only improve rehabilitation service for older people to the extent that AgeTech is affordable to people who need it, and the health systems that care for them. Developing low-cost AgeTech for rehabilitation will improve access and cost to ensure widespread uptake of rehabilitative care. Rehabilitation services are among the health services most disrupted by the COVID-19 pandemic and will play an important role in restoring function among people acutely affected by COVID-19 infections.

Find Out More

Canadian Agency for Drugs and Technology in Health (CADTH): Health Technology Update: Aging-in-place.

CADTH is an independent, not-for-profit organization responsible for providing Canada's health care decision-makers with objective evidence to help make informed decisions about the optimal use of drugs and medical devices in our health care system.

CADTH: Horizon Scans: High-level summaries of a new or emerging health technology likely to have a significant impact on the delivery of health care in Canada.

Cochrane Database of Systematic Reviews: Browse by review group.

The Cochrane Library is a collection of databases that contain high-quality, independent evidence to inform health care decision-making.

International Federation of Robotics (IFR): Free downloads on Industrial and Service Robots.

The International Federation of Robotics, established as a non-profit in 1987 connects the world of robotics around the globe.

World Health Organization: Rehabilitation in Health Systems.

Rehabilitation in health systems provides recommendations for Member States and other relevant stakeholders to strengthen and expand the availability of quality rehabilitation services. Currently, there is a significant unmet need for rehabilitation services and it is frequently undervalued in the health system. As populations age and the prevalence of noncommunicable diseases and injuries increases, and the demand for rehabilitation grows, strengthening rehabilitation in health systems becomes ever more paramount.

World Health Organization: Assistive devices and technologies.

World Health Organization: Fact sheets: Rehabilitation.

WHO provides support for assistive technology enabling people to live healthy, productive, independent, dignified lives and to participate in education, the labor market and civic life. Assistive technology reduces the need for formal health and support services, LTC and the work of caregivers. Without assistive technology, people are often excluded, isolated and locked into poverty, thereby increasing the impact of disease and disability on a person, their family and society.

World Physiotherapy: Global body for 121 physiotherapy member organisations, representing 625,000 physiotherapists.

References

Agree, E. (2014). The potential for technology to enhance independence for those aging with a disability. *Disability Health Journal, 7*(1 Suppl), S33-9. https://doi.org/10.1016/j.dhjo.2013.09.004

Alberta Health Services. (n.d.) I CAN Centre for Assistive Technology. https://www.albertahealthservices.ca/assets/programs/ps-5789-general-info-at-services.pdf

Alberta Health Services. (2013, November 14). *Glenrose-NAIT touch-screen invention makes rehabilitation fun.* https://www.albertahealthservices.ca/news/releases/2013/Page9366.aspx

Alberta Health Services (2018) Glenrose Rehabilitation Hospital Strategic Plan 2018–2023. https://www.albertahealthservices.ca/assets/hospitals/grh/grh-about-us-strategic-directions-brochure.pdf

Ali Express (n.d.). Robotic Hand Glove. https://www.aliexpress.com/w/wholesale-robotic-hand-glove.html

Alingh, J. F., Fleerkotte, B. M., Groen, B. E., Rietman, J. S., Weerdesteryn, V., van Asseldonk, E. H. F., Geurts, A. C. H., & Buurke, J. H. (2021) Effect of assist-as-needed robotic gait training on the gait pattern post stroke: A randomized controlled trial. *Journal of NeuroEngineering and Rehabilitation, 18*, Article no. 26. https://doi.org/10.1186/s12984-020-00800-4

Alqahtani, S., Joseph, J., Dicianno, B., Layton, N. A., Toro, M. L., Ferretti, E., Tuakli-Wosornu, Y. A., Chhabra, H., Neyedli, H., Lopes, C. R., Alqahtani, M. M., Van de Vliet, P., Kumagaya, S.-I., Kim, J.-B., McKinney, V., Yang, Y.-S., Goldberg, M., & Cooper, R. (2021). Stakeholder perspectives on research and development priorities for mobility assistive-technology: A literature review. *Disability and Rehabilitation: Assistive Technology, 16*(4), 362–376. https://doi.org/10.1080/17483107.2019.1650300

Alshehri, A. M. (2019). Stroke in atrial fibrillation: Review of risk stratification and preventive therapy. *Journal of Family & Community Medicine, 26*(2), 92–7. https://doi.org/10.4103/jfcm.JFCM_99_18

Alwardat, M., Etoom, M., Al Dajah, S., Schirinzi, T., Di Lazzaro, G., Simibaldi, P., Baigio Mercui, N., & Pisani, A. (2018). Effectiveness of robot-assisted gait training on motor impairments in people with Parkinson's disease: A systematic review and meta-analysis. *International Journal of Rehabilitation Research. Internationale Zeitschrift fur Rehabilitationsforschung. Revue internationale de recherches de readaptation, 41*(4), 287–6. https://doi.org/10.1097/MRR.0000000000000312

Alzheimer Society of Canada (2022). *Tracking devices.* https://alzheimer.ca/en/help-support/im-caring-person-living-dementia/ensuring-safety-security/locating-devices

Atashzar, S. F., Carriere, J., & Tavakoli, M. (2021). Review: How can intelligent robots and smart mechatronic modules facilitate remote assessment, assistance, and rehabilitation for isolated adults with neuro-musculoskeletal conditions? *Frontiers in Robotics and AI, 8*, 610529. https://doi.org/10.3389/frobt.2021.610529

Aukstakalnis, S., & Blatner, D. (1992). *Silicon mirage: The art and science of virtual reality.* Peachpit Press.

Backus, D., Winchester, P., & Tefertiller, C. (2010). Translating research into clinical practice: Integrating robotics into neurorehabilitation for stroke survivors. *Topics Stroke Rehabilitation, 17*(5), 362–370. https://doi.org/10.1310/tsr1705-362

Baniqued, P., Stanyer, E. C., Awais, M., Alazmani, A., Jackson, A. E., Mon-Williams, M. A., Mushtaq, F., & Holt, R. J. (2021). Brain-computer interface robotics for hand rehabilitation after stroke: A systematic review. *Journal of NeuroEngineering and Rehabilitation, 18*(1), Article no. 15. https://doi.org/10.1186/s12984-021-00820-8

Bass, D. (2014). A program that turns rehab into a game. *Bloomberg Business Week, 4378*, 36–37.

Beauchamp, M. K., Joshi, D., McMillan, J., Oz, U. E., Griffith, L. E., Basta, N. E., Kirkland, S., Wolfson, C., Raina, P., Canadian Longitudinal Study on Aging (CLSA) Team. (2022). Assessment of functional mobility after COVID-19 in adults aged 50 years or older in the Canadian Longitudinal Study on Aging. *JAMA Network Open, 5*(1), e2146168. https://doi.org/10.1001/jamanetworkopen.2021.46168

Bedaf, S., Gelderblom, G. J., & De Witte, L. (2015). Overview and categorization of robots supporting independent living of elderly people: What activities do they support and how far have they developed. *Assistive Technology, 27*(2), 88–100. https://doi.org/10.1080/10400435.2014.978916

Benson, I., Hart, K., Tussler, D., & van Middendorp, J. J. (2016). Lower-limb exoskeletons for individuals with chronic spinal cord injury: findings from a feasibility study. *Clinical rehabilitation*, 30(1), 73–84. https://doi.org/10.1177/0269215515575166

Bishop, L., Omofuma, I., Stein, J., Agrawal, S., & Quinn, L. (2020). Treadmill-based locomotor training with robotic pelvic assist and visual feedback: A FEASIBILITY STUDY. *Journal of Neurologic Physical Therapy: JNPT, 44*(3), 205–213. https://doi.org/10.1097/NPT.0000000000000317

Cameron, I. D., Crotty, M., & Kurrle, S. E. (2021). The future of rehabilitation for older Australians. *Medical Journal of Australia, 215*(4), 169–170. https://doi.org/10.5694/mja2.51184

Cameron, I. D., Fairhall, N., Langron, C., Lockwood, C., Monaghan, N., Aggar, C., Sherrington, C., Lord, S. R., & Kurrie, S., E. (2013). A multifactorial interdisciplinary intervention reduces frailty in older people: Randomized trial. *British Medical Journal, 11*, Article no. 65. https://doi.org/10.1186/1741-7015-11-65

Canadian Agency for Drugs and Technology in Health (CADTH). (2022). *10 trends shaping the future of health care in Canada* [Webinar]. https://youtu.be/32t_40pwU-Y

Canadian Institute for Health Information (CIHI). (2011). *Seniors and the health care system: What is the impact of multiple chronic conditions?* https://secure.cihi.ca/estore/productFamily.htm?locale=en&pf=PFC1575 https://secure.cihi.ca/free_products/air-chronic_disease_aib_en.pdf

Canadian Institute for Health Information (CIHI). (2021). QuickStats: Demographic characteristics of inpatient rehabilitation clients. CIHI. https://www.cihi.ca/en/demographic-characteristics-of-inpatient-rehabilitation-clients

CBC Edmonton. (2013, November 14). *Huge touchscreens make rehab a game for Glenrose patients.* https://www.cbc.ca/news/canada/edmonton/huge-touchscreens-make-rehab-a-game-for-glenrose-patients-1.2427156

Cecchi, F., Germanotta, M., Macchi, C., Montesano, A., Galeri, S., Diverio, M., Falsini, C., Martini, M., Mosca, R., Langone, E., Papadopoulou, D., Carrozza, M. C., & Aprile, I. (2021). Age is negatively associated with upper limb recovery after conventional but not robotic rehabilitation in patients with stroke: A secondary analysis of a randomized-controlled trial. *Journal of Neurology, 268*(2), 474–483.

Centre for Aging + Brain Health Innovation (CABHI). (2022). *jDome BikeAround: A virtual cycling program to reduce falls.* https://www.cabhi.com/completed-project-summaries/fall-prevention-program-using-the-jdome-bikearound-for-high-risk-residents-living-in-long-term-care/

Cieza, A., Causey, K., Kamenov, K., Hanson, S. W., Chatterji, S., & Vos, T. (2020). Global estimates of the need for rehabilitation based on the Global Burden of Disease study 2019: A systematic analysis for the Global Burden of Disease Study 2019. *The Lancet, 396*(10267), 2006–2017. https://doi.org/10.1016/s0140-6736(20)32340-0

Conference Board of Canada. (2017). *The market profile of physiotherapists in Canada.* https://physiotherapy.ca/sites/default/files/8695_profile-of-physiotherapists-in-canada_br.pdf

Cooper, M., & Keating, D. (1996). Implications of the emerging home systems technologies for rehabilitation. *Medical Engineering and Physics, 18*(3), 176–180. https://doi.org/10.1016/1350-4533(95)00071-2

Crea, S., Cempini, M., Mazzoleni, S., Carrozza, M., Posteraro, F., & Vitiello, N. (2017). Phase-II Clinical validation of a powered exoskeleton for the treatment of elbow spasticity. *Frontiers in Neuroscience., 11*, 261. https://doi.org/10.3389/fnins.2017.00261

Crocker, T., Forster, A., Young, J., Brown, L., & Ozer, S, Smith, J., Green, J., Hardy, J., Burns, E., Glidewell, E., & Greenwood, D. C. (2013). Physical rehabilitation for older people in long-term care. *Cochrane Database of Systematic Reviews.* https://doi.org/10.1002/14651858.CD004294.pub3

Darcy, S., Green, J., & Maxwell, H. (2017). I've got a mobile phone too! Hard and soft assistive technology customization and supportive call centres for people with disability. *Disability and Rehabilitation. Assistive Technology, 12*(4), 341–351. https://doi.org/10.3109/17483107.2016.1167260

De Luca, R., Maresca, G., Balletta, T., Cannavò, A., Leonardi, S., Latella, D., Maggio, M. G., Portaro, S., Naro, A., & Calabrò, R. S. (2020). Does overground robotic gait training improve non-motor outcomes in patients with chronic stroke? Findings from a pilot study. *Journal of Clinical Neuroscience: Official Journal of the Neurosurgical Society of Australasia, 81*, 240–5. https://doi.org/10.1016/j.jocn.2020.09.070

DHgate. (n.d.) Rehabilitation Robot Glove Hand Rehabilitation Device for Stroke Hemiplegia hand Function Recovery Hand Finger Trainer. https://www.dhgate.com/product/rehabilitation-robot-glove-hand-rehabilitation/735464370.html?f=bm%7CGMC%7C%7Bifdyn:dyn%7D%7Bifpla:pla%7D%7Bifdbm:DBM%7D%7C%7Bcampaignid%7D%7C%7Badgroupid%7D%7C735464370%7C%7Btargetid%7D%7C018007019%7CUS%7Cfengxing_beauty%7C%7Bdevice%7D%7C2%7C%7Bgclid%7D%7C&utm_source=%7Bifdyn:dyn%7D%7Bifpla:pla%7D%7Bifdbm:DBM%7D&utm_medium=GMC&utm_campaign=fengxing_beauty&utm_term=735464370&abVersion=538

Direct Supply. (2023). *Jintronix Therapy System*. https://store.directsupply.com/Product/jintronix-therapy-system-10630373?quantity=1&

Dixon, J. (2020). *Analytical studies: Methods and references how to build a robots!* [Database]. Economic Analysis Division, Statistics Canada. https://www150.statcan.gc.ca/n1/pub/11-633-x/11-633-x2020004-eng.htm

Dockx, K., Bekkers, E. M. J., Van den Bergh, V., Ginis, P., Rochester, L., Hausdorff, J. M., Mirelman A., & Nieuwboer A. (2016). Virtual reality for rehabilitation in Parkinson's disease. *Cochrane Database of Systematic Reviews*, (12), Article no. CD010760. https://doi.org/10.1002/14651858.CD010760.pub2

Dyer, S. M., Standfield, L. B., Fairhall, N., Cameron, I. D., Gresham, M., Brodaty, H., & Crotty, M. (2020). Supporting community-dwelling older people with cognitive impairment to stay at home: A modelled cost analysis. *Australasian Journal of Ageing, 39*, e506-14. https://doi.org/10.1111/ajag.12818

Erfurt-Cooper, P., & Cooper, M. (2009). Technology and economics of health and wellness spa Tourism. In *Health and Wellness Tourism: Spas and Hot Springs* (pp. 181–206). View Publications.

Federici, S., & Borsci, S. (2016). Providing assistive technology in Italy: The perceived delivery process quality as affecting abandonment. *Disability and Rehabilitation. Assistive Technology, 11*(1), 22–31. https://doi.org/10.3109/17483107.2014.930191

Furnari, A., Calabrò, R. S., De Cola, M. C., Bartolo, M., Castelli, A., Mapelli, A., Buttacchio, G., Farini, E., Bramanti, P., & Casale, R. (2017). Robotic-assisted gait training in Parkinson's disease: A three-month follow-up randomized clinical trial. *International Journal of Neuroscience, 127*(11), 996–1004. https://doi.org/10.1080/00207454.2017.1288623

Fung, J., & Keshner, E. A. (2017). The quest to apply VR technology to rehabilitation: Tribulations and treasures. *Journal of Vestibular Research, 27*(1), 1–5. https://doi.org/10.3233/VES-170610

Gélinas-Bronsard, D., Mortenson, W. B., Ahmed, S., Guay, C., & Auger, C. (2019). Co-construction of an Internet-based intervention for older assistive technology users and their family caregivers: Stakeholders' perceptions. *Disability and Rehabilitation. Assistive Technology, 14*(6), 602–611. https://doi.org/10.1080/17483107.2018.1499138

Góngora Alonso, S., Hamrioui, S., de la Torre Díez, I., Motta Cruz, E., López-Coronado, M., & Franco, M. (2019). Social robots for people with aging and dementia: A systematic review of literature. *Telemedicine Journal and E-Health: the Official Journal of the American Telemedicine Association, 25*(7), 533–540. https://doi.org/10.1089/tmj.2018.0051

Hajat, C., & Stein, E. (2018). The global burden of multiple chronic conditions: A narrative review. *Preventive Medicine Reports, 12*, 284–293. https://doi.org/10.1016/j.pmedr.2018.10.008

Hayward, K. S., Kramer, S. F., Thijs, V., Ratcliffe, J., Ward, N. S., Churilov, L., Jolliffee, L., Corbett, D., Cloud, G., Kaffenberger, T., Brodtmann, A., Bernhardt, J., & Lannin, N. A. (2019). A systematic review protocol of timing, efficacy and cost effectiveness of upper limb therapy for motor recovery post-stroke. *Systematic Review, 8*, Article no. 187. https://doi.org/10.1186/s13643-019-1093-6

Hopkins, R. O., Suchyta, M. R., Kamdar, B. B., Darowski, E., Jackson, J. C., & Needham, D. M. (2017). Instrumental activities of daily living after critical illness: A systematic review. *Annals of the American Thoracic Society, 14*(8), 1332–1343. https://doi.org/10.1513/AnnalsATS.201701-059SR

International Federation of Robotics (IFR). (2021). *Service robots: Definition*. https://ifr.org/

International Spa Association Foundation (ISPA). (2022). Now you have the advantage. 2022 ISPA U.S. spa industry study. https://experienceispa.com/wp-content/uploads/2022/08/2022-Industry-Study-FINAL.pdf

Kerkhof, Y. J. F., Graff, M. J. L., Bergsma, A., de Vocht, H. H. M., & Dröes, R.-M. (2016). Better self-management and meaningful activities thanks to tablets? Development of a person-centered program to support people with mild dementia and their carers through use of hand-held touch screen devices. *International Psychogeriatrics, 28*(11), 1917–1929. https://doi.org/10.1017/S1041610216001071

Koehn, S., Baumbusch, J., Reid, R. C., & Li, N. (2018). 'It's like chicken talking to ducks' and other challenges to families of Chinese immigrant older adults in long-term residential care. *Journal of Family Nursing, 24*(2), 156–183. https://doi.org/10.1177/1074840718774068

Kouroupetroglou, G., Pino, A., & Riga, P. A. (2017). Methodological approach for designing and developing web-based inventories of mobile assistive technology applications. *Multimedia Tools and Applications, 76*, 5347–5366. https://doi.org/10.1007/s11042-016-3822-3

Krishnaswami, A., Beavers, C., Dorsch, M. P., Dodson, J. A., Masterson Creber, R., Kitsiou, S., Goyal, C., Alexander, K. P., Batsis, J. A., Turakhia, M. P., Forman, D. E., Bernacki, G. M., Kirkpatrick, J. N., Orr, N. M., Paterson, E. D., Rich, M. W., Freeman, A. M., & Bhavnani, S. P. (2020). Gerotechnology for older adults with cardiovascular diseases: JACC state-of-the-art review. *Journal of the American College of Cardiology, 76*(22), 2650–2670. https://doi.org/10.1016/j.jacc.2020.09.606

Kwakkel, G., van Peppen, R., Wagenaar, R. C., Wood Dauphinee, S., Richards, C., Ashburn, A., Miller, K., Lincoln, N., Partridge, C., Wellwood, I., & Langhorne, P. (2004). Effects of augmented exercise therapy time after stroke: A meta-analysis. *Stroke, 35*(11), 2529–2539. https://doi.org/10.1161/01.STR.0000143153.76460.7d

Lamuzi. (n.d.). *Rehabilitation robot glove for stroke hemiplegia hand*.

Landry, M. D., Jaglal, S., Wodchis, W., Raman, J., & Cott, C. A. (2007). Analysis of factors affecting demand for rehabilitation services in Ontario, Canada: A health policy perspective. *Disability and Rehabilitation, 30*(24), 1837–1847. https://doi.org/10.1080/09638280701688078

Lansley, P., McCreadie, C., & Tinker, A. (2004). Can adapting the homes of older people and providing assistive technology pay its way? *Age & Ageing, 33*(6), 571–576. https://doi.org/10.1093/ageing/afh190

Laut, J., Porfiri, M., & Raghavan, P. (2016). The present and future of robotic technology in rehabilitation. *Current Physical Medicine Rehabilitation Reports, 4*(4), 312–319. https://doi.org/10.1007/s40141-016-0139-0

Laver, K. E., Adey-Wakeling, Z., Crotty, M., Lannin, N. A., George, S., & Sherrington, C. (2020). Telerehabilitation services for stroke. *Cochrane Database of Systematic Reviews*, (1). Article no. CD010255. https://doi.org/10.1002/14651858.CD010255.pub3

Leocani, L., Diserens, K., Moccia, M., & Caltagirone, C. (2020). Disability through COVID-19 pandemic: Neurorehabilitation cannot wait. *European Journal of Neurology, 27*(9), e50-1. https://doi.org/10.1111/ene.14320

Lo, K., Stephenson, M., & Lockwood, C. (2019). The economic cost of robotic rehabilitation for adult stroke patients: A systematic review. *JBI Database of Systematic Reviews and Implementation Reports, 17*(4), 520–547. https://doi.org/10.11124/JBISRIR-2017-003896

Lohr, K.N. (1997). Improving health care outcomes through geriatric rehabilitation: Conference summary. Medical care *35*(6),121–130.

Ludica Health. (2024). *Video game therapy app engages patients in enjoyable, affordable and convenient exercise for better balance, strength, stamina and brain health.* https://ludicahealth.com/ludica-health-launches-ludofit-pro-an-interactive-therapy-exercise-solution-for-outpatient-clinics-and-home-health/

Ludica Health. (n.d.). LudoFit PRO. *Evidence-based interactive therapy exercise for balance, strength, endurance and more.* https://ludicahealth.com/ludofitpro/

Ma, B., Yang, J., Yuet Wong, F. K., Ching Wong, A. K., Ma, T., Meng, J., Zhao, Y., Wang, Y., & Lu, Q. (2023). Artificial intelligence in elderly healthcare: A scoping review. *Ageing Research Reviews., 83*, 101808. https://doi.org/10.1016/j.arr.2022.101808

MacNeil, M., Koch, M., Kuspinar, A., Juzwishin, D., Lehoux, P., & Stolee, P. (2019). Enabling health technology innovation in Canada: Barriers and facilitators in policy and regulatory processes. *Health Policy, 123*(2), 203–214. https://doi.org/10.1016/j.healthpol.2018.09.018

Marchildon, G., & Peckham, A. (2019). *Assistive devices coverage: Ontario compared to other high-income jurisdictions.* North American Observatory on Health Systems & Policies. https://ihpme.utoronto.ca/wp-content/uploads/2018/10/NAO-Rapid-Review-5_EN.pdf

Marengoni, A., Angleman, S., Melis, R., Mangialasche, F., Karp, A., Garmen, A., Meinow, B., & Fratiglioni, L. (2011). Aging with multimorbidity: A systematic review of the literature. *Ageing Research Reviews, 10*(4), 430–439. https://doi.org/10.1016/j.arr.2011.03.003

Martel, D., Lauzé, M., Agnoux, A., Fruteau de Laclos, L., Daoust, R., Émond, M., Sirois, M.-J., & Aubertin-Leheudre, M. (2018). Comparing the effects of a home-based exercise program using a gerontechnology to a community-based group exercise program on functional capacities in older adults after a minor injury. *Experimental Gerontology, 108*, 41–7. https://doi.org/10.1016/j.exger.2018.03.016

Martín-Lesende, I., Orruño, E., Bilbao, A., Vergara, I., Cairo, M. C., Bayón, J. C., Reviriego, E., Romo, M. I., Larrañaga, J., Asua, J., Abad, R., & Recalde, E. (2013). Impact of telemonitoring home care patients with heart failure or chronic lung disease from primary care on healthcare resource use (the TELBIL study randomised controlled trial). *BMC Health Services Research, 13*, 118. https://doi.org/10.1186/1472-6963-13-118

Martín, I., Orruño, E., Mateos, M., Recalde, E., Asua, J., Reviriego, E., & Bayón, J. C. (2017). Telemonitoring in-home complex chronic patients from primary care in routine clinical practice: Impact on healthcare resources use. *European Journal of General Practice*, 23(1), 136–143. https://doi.org/10.1080/13814788.2017.1306516

Mattison, C. A., Wilson, M. G., Wang, R. H., & Waddell, K. (2020). Enhancing equitable access to assistive technologies in Canada: Insights from citizens and stakeholders. *Canadian Journal on Aging, 39*(1), 69–88. https://doi.org/10.1017/S0714980819000187

McHugh, G., & Swain, I. D. (2014). A comparison between reported therapy staffing levels and the department of health therapy staffing guidelines for stroke rehabilitation: A national survey. *BMC Health Services Research, 14*, 216. https://doi.org/10.1186/1472-6963-14-216

Means, R. (2007). Safe as houses? Ageing in place and vulnerable older people in the UK. *Social Policy and Administration, 41*(1), 65–85. https://doi.org/10.1111/j.1467-9515.2007.00539.x

Melvin J. L. (1980). Interdisciplinary and multidisciplinary activities and the ACRM. *Archives of Physical Medicine and Rehabilitation, 61*(8), 379–380. PMID. 6447491.

Ministry of Health and Long-Term Care (MOHLTC). (2014, October). *Assess & restore guideline.* https://www.health.gov.on.ca/en/pro/programs/assessrestore/docs/ar_guideline.pdf

Ministry of Health. (2022). *Assistive devices program.* Government of Ontario. https://www.ontario.ca/page/assistive-devices-program#miscellaneous%20nc

Murtagh, J. (2015). *ReWalk: Robotic exoskeleton for spinal cord injury.* (141). Canadian Agency for Drugs and Technology in Health. https://www.cadth.ca/sites/default/files/pdf/ReWalk_Robotic_Exoskeletons_for_Spinal_Cord_Injury_e.pdf

Myburg, M., Allan, E., Nalder, E., Schuurs, S., & Amsters, D. (2017). Environmental control systems—The experiences of people with spinal cord injury and the implications for prescribers. *Disability and Rehabilitation. Assistive Technology, 12*(2), 128–136. https://doi.org/10.3109/17483107.2015.1099748

Nasr, N., Mawson, S., Wright, P., Parker, J., & Mountain, G. (2016). Exploring the experiences of living with stroke through narrative: Stroke survivors' perspectives. *Global Qualitative Nursing Research, 3.* https://doi.org/10.1177/2333393616646518

O'Brien Institute of Public Health. (2022). *Rapid solutions for COVID-19 in continuing care: Virtual reverse trade show.* University of Calgary; O'Brien Institute of Public Health. https://obrieniph.ucalgary.ca/rapid-solutions-for-COVID-19-in-continuing-care

Oh, A., Patel, K., Boscardin, W. J., Max, W., Stephens, C., Ritchie, C. S., & Smith, A. K. (2019). Social support and patterns of institutionalization among older adults: A longitudinal study. *Journal of the American Geriatrics Society, 67*, 2622–2627. https://doi.org/10.1111/jgs.16184

Papi, E., Murtagh, G. M., & McGregor, A. H. (2016). Wearable technologies in osteoarthritis: A qualitative study of clinicians' preferences. *British Medical Journal Open, 6*(1), e009544. https://doi.org/10.1136/bmjopen-2015-009544

Parker, V. A., & Geron, S. M. (2007). Cultural competence in nursing homes. *Gerontology & Geriatrics Education, 28*(2), 37–54. https://doi.org/10.1300/J021v28n02_04

Peek, S. T. M., Wouters, E. J., Luijkx, K. G., & Vrijhoef, H. J. (2016). What it takes to successfully implement technology for aging in place: Focus groups with stakeholders. *Journal of Medical Internet Research, 18*(5), e98. https://doi.org/10.2196/jmir.5253

Picelli, A., Melotti, C., Origano, F., Waldner, A., Gimigliano, R., & Smania, N. (2012). Does robotic gait training improve balance in Parkinson's disease? A randomized controlled trial. *Parkinsonism & Related Disorders, 18*(8), 990–993. https://doi.org/10.1016/j.parkreldis.2012.05.010

Pillmayer, M., Scherle, N., Pforr, C., Locher, C., & Herntrei, M. (2021). Transformation processes in Germany's health resorts and spas—A three case analysis. *Annals of Leisure Research, 24*(3), 310–327. https://doi.org/10.1080/11745398.2020.1765399

Porras, D. C., Siemonsma, P., Inzelberg, R., Seilig, G., & Plotnik, M. (2018). Advantages of virtual reality in the rehabilitation of balance and gait. *Neurology, 90*(22), 1017–1025.

Quinn, W., O'Brien, E., & Springan, G. (2018). *Using telehealth to improve home-based care for older adults and family caregivers.* AARP Public Policy Institute. https://www.aarp.org/content/dam/aarp/ppi/2018/05/usingtelehealth-to-improve-home-based-care-for-older-adults-and-family-caregivers.pdf

Rehab Care Alliance. (2014). *Referral decision tree for rehabilitative care.* http://rehabcarealliance.ca/uploads/File/Toolbox/Definitions/Referral_Decision_Tree_for_Rehabilitative_Care_FINAL__Dec_11_2014_.pdf https://rehabcarealliance.ca/wp-content/uploads/2022/09/Referral_Decision_Tree_for_Rehabilitative_Care_FINAL__Dec_11_2014_.pdf

Reyes, A., Qin, P., & Brown, C. A. (2018). A standardized review of smartphone applications to promote balance for older adults. *Disability and Rehabilitation, 40*(6), 690–696. https://doi.org/10.1080/09638288.2016.1250124

Robinson, L., Gibson, G., Kingston, A., Newton, L., Pritchard, G., Finch, T., & Brittain, K. (2013). Assistive technologies in caring for the oldest old: A review of current practice and future directions. *Aging and Health, 9*(4), 365–375. https://doi.org/10.2217/ahe.13.35

Sale, P., Franceschini, M., Waldner, A., & Hesse, S. (2012). Use of the robot assisted gait therapy in rehabilitation of patients with stroke and spinal cord injury. *European Journal of Physical and Rehabilitation Medicine, 48*(1), 111–121. PMID: 22543557.

Saglia, J. A., Tsagarakis, N. G., Dai, J. S., & Caldwell, D. G. (2012). Control strategies for patient-assisted training using the ankle rehabilitation robot (ARBOT). *IEEE/ASME Transactions on Mechatronics*, 18(6), 1799–1808.

Saposnik, G., Levin, M., & Outcome Research Canada Working Group. (2011). Virtual reality in stroke rehabilitation: A meta-analysis and implications for clinicians. *Stroke, 42*(5), 1380–1386. https://doi.org/10.1161/STROKEAHA.110.605451

Schultheis, M. T., & Rizzo, A. A. (2001). The application of virtual reality technology in rehabilitation. *Rehabilitation Psychology, 46*(3), 296–311. https://doi.org/10.1037/0090-5550.46.3.296

Seaborn, K., Pennefather, P., & Fels, D. I. (2016). "Learn what we're going through": Attitudes of older powered chair users towards mixed reality games that involve power mobility. *Universal Access in the Information Society, 15*(4), 699–711. https://doi.org/10.1007/s10209-015-0450-z

Shahabi, S., Rezapour, A., & Arabloo, J. (2021). Economic evaluations of physical rehabilitation interventions in older adults with hip and/or knee osteoarthritis: A systematic review. *European Journal of Physiotherapy, 23*(3), 185–195. https://doi.org/10.1080/21679169.2019.1672785

Sheehy, L., Taillon-Hobson, A., Sveistrup, H., Bilodeau, M., Yang, C., Welch, V., Hossain, A., & Finestone, H. (2019). Home-based virtual reality training after discharge from hospital-based stroke rehabilitation: A parallel randomized feasibility trial. *Trials, 20*(1), Article no. 333. https://doi.org/10.1186/s13063-019-3438-9

Sheridan, P. E., Mair, C. A., & Quiñones, A. R. (2019). Associations between prevalent multimorbidity combinations and prospective disability and self-rated health among older adults in Europe. *BMC Geriatrics, 19*, Article no. 198. https://doi.org/10.1186/s12877-019-1214-z

Shi, P., Zheng, J., & Yu, H. (2018). A virtual reality rehabilitation training system based on upper limb exoskeleton robot. [Conference presentation]. In *10th International conference on intelligent human-machine systems and cybernetics*, 2018, August 25–26, Hangzhou, China. https://doi.org/10.1109/IHMSC.2018.00058

Stefanov, D. H., Bien, Z., & Bang, W.-C. (2004). The smart house for older persons and persons with physical disabilities: Structure, technology arrangements, and perspectives. *IEEE Transactions on Neural Systems Rehabilitation Engineering, 12*(2), 228–250. https://doi.org/10.1109/TNSRE.2004.828423

Stier-Jarmer, M., Kus, S., Frisch, D., Sabariego, C., & Schuh, A. (2015). Health resort medicine in non-musculoskeletal disorders: Is there evidence of its effectiveness? *International Journal of Biometeorology, 59*(10), 1523–1544. https://doi.org/10.1007/s00484-015-0953-6

Sund, T. (2017). *Assistive technology in Norway- a part of a larger system.* https://www.nav.no/_/attachment/inline/7b119b1c-fe72-488a-a1ef-be424e72faff:c52b8c6ee759299749538a6fd0554d1efa695abf/assistive-technology-in-norway-170217v2.pdf

Swain, S., Sarmanova, A., Coupland, C., Doherty, M., & Zhang, W. (2020). Comorbidities in osteoarthritis: A systematic review and meta-analysis of observational studies. *Arthritis Care & Research, 72*(7), 991. https://doi.org/10.1002/acr.24008

Takano, E., Ozaki, K., Satoh, K., Kawamura, K., Maltais, M., & Kondo, I. (2020). Effects of a balance exercise assist robot on older patients with hip fracture: A preliminary study. *Journal of Medical and Biological Engineering, 40*(6), 783–789. https://doi.org/10.1007/s40846-020-00568-x]

Teulieres, M., Partner, Tilley, J., Bolz, L., Ludwig-Dehm, P. M., & Wägner, S. (2019). *Industrial robotics: Insights into the sector's future growth dynamics.* McKinsey & Company. https://www.mckinsey.com/~/media/mckinsey/industries/advanced%20electronics/our%20insights/growth%20dynamics%20in%20industrial%20robotics/industrial-robotics-insights-into-the-sectors-future-growth-dynamics.ashx

The World Bank. (2019). *Life expectancy at birth, total (years), Canada.* The World Bank. https://data.worldbank.org/indicator/SP.DYN.LE00.IN?locations=CA

Tieri, G., Morone, G., Paolucci, S., & Iosa, M. (2018). Virtual reality in cognitive and motor rehabilitation: Facts, fiction and fallacies. *Expert Review of Medical Devices, 15*(2), 107–117. https://doi.org/10.1080/17434440.2018

Tomšič, M., Domajnko, B., & Zajc, M. (2018). The use of assistive technologies after stroke is debunking the myths about the elderly. *Topics in Stroke Rehabilitation, 25*(1), 28–36. https://doi.org/10.1080/10749357.2017.1376845

Venketasubramanian, N. (2020). Stroke care services in Singapore during COVID-19 pandemic—A national perspective. *Frontiers in Neurology, 11*, 780. https://doi.org/10.3389/fneur.2020.00780

van Balen, R., Gordon, A. L., Schols, J. M., Drewes, Y. M., & Achterberg, W. P. (2019). What is geriatric rehabilitation and how should it be organized? A delphi study aimed at reaching European consensus. European Geriatric Medicine, 10, 977–987.

Wang, R. H., Sudhama, A., Begum, M., Huq, R., & Mihailidis, A. (2017). Robots to assist daily activities: Views of older adults with Alzheimer's disease and their caregivers. *International Psychogeriatrics, 29*(1), 67–79. https://doi.org/10.1017/S1041610216001435

Wang, X., Hunter, D. J., Vesentini, G., Pozzobon, D., & Ferreira, M. L. (2019). Technology-assisted rehabilitation following total knee or hip replacement for people with osteoarthritis: A systematic review and meta-analysis. *BMC Musculoskeletal Disorders, 20*(1), 1–17.

Wilson, D. J., Mitchell, J. M., Kemp, B. J., Adkins, R. H., & Mann, W. (2009). Effects of assistive technology on functional decline in people aging with a disability. *Assistive Technology, 21*(4), 208–217. https://doi.org/10.1080/10400430903246068

World Health Organization (WHO); The World Bank. (2011). *World report on disability.* World Health Organization. https://www.who.int/teams/noncommunicable-diseases/sensory-functions-disability-and-rehabilitation/world-report-on-disability

World Health Organization (WHO). (2017). *Rehabilitation in health systems.* Licence: CC BY-NC-SA 3.0 IGO. https://www.who.int/publications/i/item/9789241549974

World Health Organization (WHO). (2020). *Physical activity: Key facts.* https://www.who.int/news-room/fact-sheets/detail/physical-activity

World Health Organization (2021a). *Assistive devices and technologies. Key facts.* https://www.who.int/news-room/fact-sheets/detail/assistive-technology

World Health Organization (2021b). *Fact sheets: Rehabilitation.* https://www.who.int/news-room/fact-sheets/detail/rehabilitation

World Health Organization (WHO) and United Nations Children's Fund (UNICEF). (2022). *Global report on assistive technology.* World Health Organization and the United Nations Children's Fund (UNICEF). https://www.who.int/publications/i/item/9789240049451

World Physiotherapy. (2022). Profile of the global professions. How many practising physiotherapists per 10,000 population? https://www.world.physio/membership/profession-profile

Yurkewich, A. (2015). *Design and Control of Robotic Systems for Lower Limb Stroke Rehabilitation* Electronic Thesis and Dissertation Repository. 3077. https://ir.lib.uwo.ca/etd/3077

Zampolini, M., Todeschini, E., Hermens, H., Ilsbroukx, S., Macellari, V., Magni, R., & Giacomozzi, C. (2008). Tele-rehabilitation: present and future. Annali dell'Istituto superiore di sanita, *44*(2), 125–134.

Primary Care and Care Coordination 7

The Challenge

Older adults are frequent users of primary care but are also at increased risk of health care related adverse events and resulting poor health outcomes. Despite having what are perceived as key central roles in gatekeeping or care coordination, primary care faces a number of challenges in performing these roles, including time and resource constraints, and lack of information for clinical care and quality improvement. Transitions in care are common for older patients who have multiple, complex health issues and thus require services from multiple providers and settings, but health care settings often operate independently as "silos." As older patients transition from one care setting to another, they often experience challenges related to inadequate communication or information transfer (Fig. 7.1).

Persona and Scenario—Mariam

Mariam is an 86-year-old woman living alone in her own home. She has been living alone since her husband passed away three years ago. Mariam had been the primary caregiver for her husband who had required considerable assistance with his daily activities. Mariam was proud that she had been able to care for her husband at home, but her own health declined during this period. She was recently discharged from hospital, to which she had been admitted for investigation of cardiac arrythmia. During her hospital stay, she acquired a Clostridium difficile infection that caused symptoms of delirium and increased the length of her hospital stay. On her return home, she was referred to home care to assess her home living situation and to provide rehabilitation therapy to recover mobility lost due to deconditioning while in hospital. Her family physician was not advised of

D. Juzwishin et al., *AgeTech Innovations in Healthcare for Older Adults*, Synthesis Lectures on Technology and Health, https://doi.org/10.1007/978-3-031-86885-6_7

Fig. 7.1 Health care team sharing information. *Credit* Shutterstock: MZStoc, https://www.shutterstock.com/image-photo/group-multiethnic-medical-professionals-including-doctors-2295175091

the referral or the results of the case manager's assessment. Mariam will also be followed regularly by a cardiac specialist. Mariam wishes to remain in her own home but is anxious about how she will be able to manage. Her daughter Asha can provide some support but works and has a young family of her own. Asha is worried about her mother's safety as Mariam has been showing some signs of cognitive decline. One of Mariam's challenges is getting to her primary care and specialist appointments, as she does not drive and is uncomfortable taking public transit, including the para-transit van.

Background

The focus of this chapter is on the role of technology in primary care for older patients and in the coordination of their care between care providers and care settings.

Primary Care

- Primary care is: "a key process in the health system that supports first-contact, accessible, continued, comprehensive and coordinated patient-focused care" [World Health Organization (WHO) and the United Nations Children's Fund (UNICEF), 2020, p. xiii].
- Primary care plays a key role in prevention, health monitoring and promotion and in care coordination.
- Four core functions of primary care (Jimenez et al., 2021):
  - first contact
  - comprehensiveness
  - coordination
  - continuity
- Older adults are frequent users of primary care and are at increased risk of health care related adverse events.
- Systems of primary care are diverse and influenced by the context and country in which they have been developed.

The WHO (2020, p. xiii) defines primary care as, "a key process in the health system that supports first-contact, accessible, continued, comprehensive and coordinated patient-focused care." Similarly, Jimenez and colleagues (2021) describe four core functions of primary care—the "4Cs" of first contact, comprehensiveness, coordination and continuity.

Primary care plays a key role in prevention, health monitoring and promotion and in care coordination. Benefits of a well-developed system of primary care include improved patient and population health outcomes, continuity of care and cost containment (Kringos et al., 2015; Macinko et al., 2003; Starfield et al., 2005).

Older adults are frequent users of primary care but are also at increased risk of health care related adverse events and resulting poor health outcomes (Cooper et al., 2017). In a United Kingdom study, Cooper and colleagues (2017) found that unsafe primary care was mainly due to medication-related incidents, communication-related incidents and clinical decision-making incidents.

For older patients, particularly those with multiple health conditions, transitions between providers and care settings are common. Information sharing, particularly at points of discharge from one care setting to another is thus critical to support continuity of care and to avoid adverse events. Countries with stronger primary care systems report more positive coordination experiences (Schoen et al., 2008). However older patients, particularly those with chronic conditions, often experience poorly coordinated care (Stolee et al., 2019; Stolee et al., 2021).

Systems of primary care are diverse and are influenced by the context and historical background of the country and jurisdiction in which they have been developed (Kringos et al., 2015). Kringos and colleagues (2015) have identified three types of primary care organization (1) the public hierarchical normative model in which primary care has a central role in the health care system, and is run by the state; (2) the professional hierarchical gatekeeper model, with general practitioners in a pre-eminent gatekeeping role; and (3) the free professional non-hierarchical model, in which primary care is organized by health professionals independently or with limited regulation.

Primary care is often described as having a "gatekeeper" role, controlling referrals to specialists (Forrest, 2003) and other services, although this is not the case in all systems (Schmalstieg-Bahr et al., 2021). Bodenheimer and colleagues (1999, 2014) have argued for a role as coordinators rather than as gatekeepers, to integrate primary and specialty care.

Care Coordination

- Care coordination role for primary care in "organizing patient care activities and sharing information among all of the participants concerned with a patient's care to achieve safer and more effective care helping the patient navigate through the health care system" [Agency for Healthcare Research and Quality (AHRQ), 2018, para. 1].
- Health care providers who are not part of primary care teams, such as home care case managers and coordinators and social workers, also play roles in care coordination.
- Transitions in care are common for older patients who have multiple, complex health issues and thus require services from multiple providers and settings.
- As older patients transition from one care setting to another, they often experience challenges related to inadequate communication or information transfer—these challenges can lead to adverse outcomes.

A care coordination role for primary care can be seen more broadly as "organizing patient care activities and sharing information among all of the participants concerned with a patient's care to achieve safer and more effective care helping the patient navigate through the health care system" (AHRQ, 2018, para. 1). The AHRQ (2018, para. 6) provides the following examples of care coordination activities:

- Establishing accountability and agreeing on responsibility.
- Communicating/sharing knowledge.
- Helping with transitions of care.
- Assessing patient needs and goals.
- Creating a proactive care plan.
- Monitoring and follow-up, including responding to changes in patients' needs.
- Supporting patients' self-management goals.

- Linking to community resources.
- Working to align resources with patient and population needs.

Health care providers who are not part of primary care teams, such as home care case managers and coordinators as well as social workers, also play roles in care coordination. These roles are often aligned with a specific health setting or coordination function, such as hospital discharge planning, or coordination of home and community care services (Fig. 7.2).

Transitions in care are common for older patients who have multiple, complex health issues and thus require services from multiple providers and settings. In most health care systems, health care settings often operate independently as 'silos' (Coleman, 2003). As older patients transition from one care setting to another, they often experience challenges related to inadequate communication or information transfer (Glenny et al., 2013). These shortcomings in communication can increase the patient's risk of readmission, medication errors, or other adverse events (Forster et al., 2003). System or patient navigators have been proposed to help patients in their interactions with the health care system and to navigate through care transitions (Kokorelias et al., 2021; Manderson et al., 2012). Multifaceted programs have also been developed specifically to aid patients through transitions; these include programs developed by Coleman et al. (2006), Naylor et al. (2004) and Jack

Fig. 7.2 Care coordination challenges

et al. (2009). These programs support patient education and information sharing, and often involve a coach or navigator role. Transition coaches or navigators perform a useful function, but the need for persons in these roles is a recognition that health care systems are not well integrated or coordinated. In addition, the issue of adequate compensation for these services has also been raised.

Opportunities for Technology in Primary Care and Care Coordination

Despite having what is perceived as a key central role in gatekeeping or care coordination, primary care faces a number of challenges in performing this role, including time and resource constraints, and lack of information for clinical care and quality improvement (Lafortune et al., 2015; Young & Nesbitt, 2017). There are growing pressures for the use of technology in primary care, driven by a population that is increasingly comfortable with the use of technology (and accompanying expectations of quick access to information) and shortages of primary care physicians (Pravia & Diaz, 2019).

There are a variety of opportunities for technologies to aid in addressing challenges in primary care settings. Young and Nesbitt (2017) have proposed a framework, see Fig. 7.3, for technology-enabled primary care that includes four domains: the body, the home, the community and the primary care clinic.

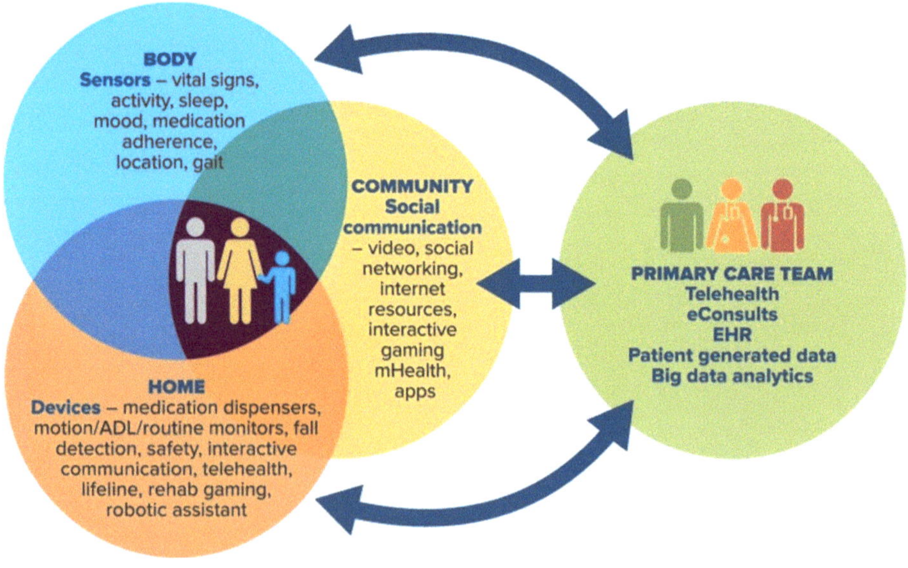

Fig. 7.3 Technology Enabled Primary Care *Image Source* Young and Nesbitt (2017)

This chapter will focus primarily on technologies that may be connected to primary care providers, specifically (1) virtual care; and (2) mHealth applications and e-visits. Electronic health records and electronic health information systems are discussed in Chap. 8. Technologies focused on the home environment are discussed in Chap. 4.

Opportunities for Technology

- Primary care faces many challenges, including time and resource constraints, and lack of information for clinical care and quality improvement.
- There are many opportunities for technologies to aid in addressing these challenges. Examples include:
  - Virtual care, including telehealth and technologies for remote monitoring.
  - mHealth applications and e-visits.
  - Electronic health research and electronic health information technology, including online patient portals and embedded risk and frailty screening (discussed in Chap. 8).
- Patients and practitioners need to be involved in the design and implementation of the technologies, in the context of effective models of care delivery that support positive relationships of patients and providers.
- AgeTech will not be a panacea for resource constraints and health system issues.

Virtual Care

Virtual care has been defined as: "any interaction between patients and/or members of their circle of care, occurring remotely, using any forms of communication or information technologies, with the aim of facilitating or maximizing the quality and effectiveness of patient care (Shaw et al., 2018, p. 609). While the terms "telehealth" and "telemedicine" are sometimes considered as overarching or umbrella categories which include applications of technology for a range of remote activities, including education and health administration (Doarn et al., 2014), these terms and "virtual care" will be used interchangeably in this chapter.

Telemedicine is expected to grow in popularity, especially in North America and Europe where it has had governmental support (Landi, 2018). Agarwal and colleagues (2020) examined the use of telemedicine for primary care in Canada and Brazil where telemedicine offers advantages given the large geographies of both countries. In Brazil, its use for clinical interactions with patients has been limited due to beliefs in the importance of the physical presence of practitioners and prohibitions on the use of teleconsultation for patient communication; telemedicine in Canada has had stronger regulatory support (Agarwal et al., 2020).

Technologies have been used for remote monitoring of vital signs, such as oxygen saturation, which can be combined with self-reported symptoms to guide care planning (Akhtar & Pham, 2015). Telehealth solutions also include services for remote consultations using the telephone or computer monitors (Akhtar & Pham, 2015).

COVID-19 Pandemic Led to Dramatic Increase in Use of Virtual Care

The need to maintain physical distancing caused by the COVID-19 pandemic led to a dramatic increase in the use of virtual care approaches (Glazier et al., 2021; Neves et al., 2021; Wong et al., 2021). Neves and colleagues (2021) examined virtual primary care during the pandemic in nine high-income countries; they found that despite successes, the increased use of virtual approaches also revealed weaknesses. Based on their review, they recommend "health policies to guide the allocation of limited resources better, greater investments to modernize infrastructure, expansion of technical training, prioritization of systems interoperability, and more support for patients with lower health and digital literacy levels" (Neves et al., 2021, p. 245). A desire for convenient, alternative means for access pre-dated the pandemic (Bayliss et al., 2008; Goldberg et al., 2020). However virtual care activities have not been in routine use by health care systems in many industrialized countries despite the availability of the technology to do so (Canadian Medical Association, 2019). Interest in virtual care may be greater among younger patients; pre-pandemic research in the United Kingdom found that older patients were less likely to choose virtual care, if other options were available (Hammersley et al., 2019).

Virtual Care has Both Advantages and Disadvantages

An obvious disadvantage of virtual care is the lack of physical hands-on examination (Breton et al., 2021). Hammersley and colleagues found somewhat less information sharing occurred in virtual visits, though efforts toward building rapport may have been greater (Hammersley et al., 2019). Older adults have also noted advantages of virtual care in terms of the convenience of not having to travel to appointments, especially in bad weather and for people who live in rural and remote areas (Elliott et al., 2021). Older adults also noted that virtual appointments served to lessen their risk of exposure to the COVID-19 virus (Elliott et al., 2021).

Shaw and colleagues (2018) led a consensus process to generate recommendations for virtual care in primary care settings; their process included patients and health system representatives. The process yielded a recommendation for leadership to embed virtual care in the primary care sector but with an emphasis on meaningful engagement of patients to ensure the technology met their needs. Rogers and colleagues (2011) undertook a qualitative investigation of the integration of "telecare" in managing long term conditions; they found that technology often reinforced a passive role for patients in their own care.

Increased health system emphasis on virtual care options raises potential concerns that a "digital divide" (Gonzales, 2016), which separated those with or without the resources to use new technologies, may exacerbate inequities in health system access. Because of the differences in technology use between older and younger generations, this has also been described as a "grey divide" (Friemel, 2014, p. 1). On the other hand, research in Canada found similar rates of adoption of virtual care by older and younger patients during the pandemic, although it should be noted that most virtual visits were done by telephone (Bhatia et al., 2021). Jones (2015) has argued that the technological divide is driven more by factors such as race, education and economic status than by age.

Virtual care has the potential to both reduce and exacerbate health system inequities. Virtual care can enhance access for persons in rural and remote communities and can be "a cost and time saver for patients" (Falk, 2021, p. 34). Falk (2021) has described that virtual care can enhance cultural safety, for example for Indigenous persons who may find virtual visits less culturally intimidating. On the other hand, Falk (2021) notes the danger that virtual care offers a means to avoid providing direct service in rural communities or for providers to stay out of Indigenous communities even when a physical presence is needed.

A reason for slow adoption of virtual care in the past is that public health insurance has often not reimbursed practitioners for virtual appointments. During the pandemic, many jurisdictions introduced fee codes to cover virtual appointments, however there was some concern that these may be rolled back after the pandemic (Falk, 2021).

Some jurisdictions are moving toward more rapid adoption of virtual care. In the United Kingdom the long-term plan of the National Health Service (NHS) is for "digital first primary care" to become an option for every patient to improve access (NHS, 2019, p. 99).

Virtual care has limitations and will not be the preferred modality of all patients or providers. The pandemic has likely proven the feasibility and value of virtual care however, and it will remain an important option for primary care practitioners. As Falk (2021, p. 3) notes, "Care is Care." Future adoption for use with older adults should be guided by their values, preferences and circumstances, and ensure equitable access to the necessary technology.

How Could These Technologies Help Mariam?

Because Mariam finds it difficult to travel to appointments, virtual appointments with primary care practitioners and specialists would be very helpful. Mariam's daughter may not be able to take her mother to in-person appointments but would more likely be able to manage helping her mother get set up with the necessary technology.

Mhealth and E-visits

The WHO defines mHealth as "medical and public health practice supported by mobile devices, such as mobile phones, patient monitoring devices, personal digital assistants (PDAs), and other wireless devices" (WHO, 2011, p. 6). Mobile apps have had a major focus on wellness and fitness, however they are increasingly used for management of complex health conditions (Ghosh et al., 2014). Matthew-Maich and colleagues (2016) completed a scoping review of mHealth technologies for managing chronic conditions in older adults. Most of the mHealth solutions identified were designed for use by both patients and health care providers. Their review found that mHealth technology use in health care for older adults was limited but increasing. Mercer and colleagues (2015) have also noted that most mHealth research has focused on younger people. mHealth devices that link to electronic health records, see for example the work of Ammenwerth et al. (2015) and Klein et al. (2015), offer advantages in sharing information with health care providers but may meet resistance from practitioners if they are seen as increasing workloads. Matthew-Maich et al. (2016) recommend that future mHealth design and development should be interdisciplinary and collaborative and follow a user-centered approach with feedback from patients, caregivers and health care providers, and with consideration of technology literacy and personal preferences. A note of caution in this regard comes from a co-design process for an mHealth app reported by Tong et al. (2022), in which system developers were unwilling or unable to provide features desired by older adults. These authors recommend ensuring that all parties involved are committed to the co-design process.

In some countries, automated systems are being used as alternatives to, or as a triage process for, in-person visits. In Sweden, there has been increased use of e-visits, including through email and text messaging via Web portals (Nilsson et al., 2021). Nilsson and colleagues (2021) studied patient experience with a chat-based automated medical history-taking service. They noted that such services have become increasingly popular in both the for-profit and non-profit sectors, despite controversies regarding reimbursement. Their study found a generally positive patient attitude toward e-visits, but a need for more extensive user experience design (Nilsson et al., 2021). In England, Miller and colleagues (2020) tested a mobile app for primary care patients to enter their symptoms and medical history, following which they were given a care recommendation. The study found that patients might have considered less urgent care options, e.g., pharmacy, had they received the recommendations prior to arriving in primary care. The app was intended for home use but tested in a primary care setting. Chatbots have been developed to support self-management. Hauser-Ulrich and her colleagues (2020) in Switzerland conducted a pilot RCT of a smartphone-based chatbot to promote self-management of chronic pain (SELMA). Study participants, middle-aged persons with chronic pain, provided positive feedback and suggestions for improvement. Additional research on chatbots should be conducted with older adult patients.

Some e-visit apps support a one-to-one video consultation with a physician. TELUS Health offers the Babylon app (https://www.telus.com/en/personal-health/my-care?CMP=van_babylon) that can be downloaded to a smartphone. In the Canadian province of Alberta, the provincial government covered the costs of the virtual visits; this was controversial as family physicians in Alberta complained that this was substandard care and that the Babylon doctors were paid more for virtual visits also offered by local family physicians (CBC News, 2020). Privacy concerns have also been raised (Rusnell, 2021). In the United Kingdom, use of Babylon's GP (General Physician) at hand service has grown dramatically. This growth has been accompanied by concerns about flaws in its AI system—it has a built-in symptom checking chatbot—and the financial strain it has placed on the NHS (Burgess & Kobie, 2019). Ada is another example of an AI-powered app. Ada, based in London and Berlin, serves as a symptom checker which can offer a follow-up text-based consultation with a doctor (O'Hear, 2017).

E-visits may be conducted asynchronously, using "store-and-forward" methods in which health information is electronically stored and then forwarded for review by a practitioner (Culmer et al., 2023). A systematic review of asynchronous telemedicine by Culmer et al. (2023) found that asynchronous approaches provide efficient means to augment, rather than replace, synchronous care. This review was not specifically focused on older patients. Recently, Judson et al. (2023) conducted a small qualitative study in San Francisco of older patients' perceptions of e-visits. They found that older patients were generally willing to try e-visits, but expressed preferences for synchronous communication as well as discomfort with using the technology.

Research evidence on electronic medical consultations in primary care is limited. A review conducted by Mold and colleagues (2019) found some positive benefits for care delivery, but a need for additional research to investigate usage, cost, quality and sustainability as well as greater efforts to address privacy issues and to ensure the systems do not increase inequities in access.

How Could These Technologies Help Mariam?

Mariam, given her difficulties with transportation to physician visits, might benefit from e-visits or mHealth apps. She would need to be given appropriate guidance and support in their use, particularly given her cognitive decline.

Technology-Enabled Primary Care and Care Coordination

Technology offers tremendous potential benefits for the development of stronger primary care and care coordination systems (WHO, 2020). To realize these benefits, patients and practitioners need to be actively involved in the design and implementation of the technologies (Bélanger et al., 2012). The technologies also need to be developed in the context of effective models of care delivery that support positive and effective relationships of patients and providers (Bauer et al., 2014). Young and Nesbitt (2017, p. 402) expressed similar sentiments in the conclusion of their review on the potential for technology to increase the capacity of the primary care system:

> A technology-enabled health care delivery model has the potential to increase the role and relevance of primary care as the integrative hub for health care, supporting the efforts of patients, promoting collaboration with specialists, facilitating connections with nursing homes and home health, and advancing population health within a framework that improves patient experience, quality and value. Achieving this will require better preparation for patients and family caregivers, more training and professional development for clinicians, more systematic deployment of the technology tools themselves, and changes in policy and reimbursement to recognize the tools' value. Simultaneously, primary care, through professional development and practice redesign, must reimagine its role, contribution, and potential within the context of a technology enabled delivery system.

"The secret sauce of primary care," Danielle Martin has stated, "is the relationship" (2017, p. 62).

Seminal Demonstration Project

E-visits can achieve greater time efficiencies for practitioners and convenience for patients (Culmer et al., 2023), however barriers remain related to their acceptability and usability by older patients (Judson et al., 2023). We propose a project to develop and implement an e-visit intervention for older primary care patients, guided by co-design processes (further discussed in Chap. 10) that actively engage older patients, their caregivers, and practitioners. Co-design processes would aim: (i) to identify and address barriers to implementation and acceptance; (ii) to integrate e-visits into primary health care practice; (iii) to focus the e-visit initiative on outcomes of interest to all parties involved; and (iv) to investigate if and how patient-provider relationships are affected.

Concluding Remarks

We conclude this chapter with a caution similar to that mentioned at the close of the chapter on LTC. AgeTech will not be a panacea for resource constraints and health system issues that constrain primary care practice and complicate coordination of care and services. We have, however, identified multiple opportunities for AgeTech to enhance primary care and care coordination for older adults, and to support patient-practitioner communication. We recommend that these technologies be considered as a key element in health system reforms aimed at improving primary care and the integration of health services for older people.

References

Agarwal, P., Kithulegoda, N., Umpierre, R., Pawlovich, J., Pfeil, J. N., D'Avila, O. P., Goncalves, M., Harzheim, E., & Ponka, D. (2020). Telemedicine in the driver's seat: New role for primary care access in Brazil and Canada: The Besrour Papers: A series on the state of family medicine in Canada and Brazil. *Canadian Family Physician/medecin De Famille Canadien, 66*(2), 104–111. PMID: 32060190.

Agency for Healthcare Research and Quality (AHRQ). (2018). *Care coordination.* [Website]. Agency for Healthcare Research and Quality. https://www.ahrq.gov/ncepcr/care/coordination.html

Akhtar, S., & Pham, T.-N. (2015). *A Canadian provider's handbook to home-based primary care.* University of Toronto.

Ammenwerth, E., Woess, S., Baumgartner, C., Fetz, B., van der Heidt, A., Kastner, P., Modre-Osprian, R., Welte, S., & Poelzl, G. (2015). Evaluation of an integrated telemonitoring surveillance system in patients with coronary heart disease. *Methods of Information in Medicine, 54*(5), 388–397. https://doi.org/10.3414/ME15-02-0002

Bauer, A. M., Thielke, S. M., Katon, W., Unützer, J., & Areán, P. (2014). Aligning health information technologies with effective service delivery models to improve chronic disease care. *Preventive Medicine, 66*, 167–172. https://doi.org/10.1016/j.ypmed.2014.06.017

Bayliss, E. A., Edwards, A. E., Steiner, J. F., & Main, D. S. (2008). Processes of care desired by elderly patients with multimorbidities. *Family Practice, 25*(4), 287–293. https://doi.org/10.1093/fampra/cmn040

Bélanger, E., Bartlett, G., Dawes, M., Rodríguez, C., & Hasson-Gidoni, I. (2012). Examining the evidence of the impact of health information technology in primary care: An argument for participatory research with health professionals and patients. *International Journal of Medical Informatics, 81*(10), 654–661. https://doi.org/10.1016/j.ijmedinf.2012.07.008

Bhatia, R.S., Chu, C., Pang, A., Tadrous, M., Stamenova, V., & Cram, P. (2021). Virtual care use before and during the COVID-19 pandemic: A repeated cross-sectional study. *CMAJ Open, 9*(1). https://doi.org/10.9778/cmajo.20200311

Bodenheimer, T., Lo, B., & Casalino, L. (1999). Primary care physicians should be coordinators, not gatekeepers. *JAMA, 281*(21), 2045–2049. https://doi.org/10.1001/jama.281.21.2045

Bodenheimer, T., Ghorob, A., Willard-Grace, R., & Grumbach, K. (2014). The 10 building blocks of high-performing primary care. *Annals of Family Medicine, 12*(2), 166–171. https://doi.org/10.1370/afm.1616

Breton, M., Sullivan, E. E., Deville-Stoetzel, N., McKinstry, D., DePuccio, M., Sriharan, A., Deslauriers, V., Dong, A., & McAlearney, A. S. (2021). Telehealth challenges during COVID-19 as reported by primary healthcare physicians in Quebec and Massachusetts. *BMC Family Practice, 22*, Article no. 192. https://doi.org/10.1186/s12875-021-01543-4

Burgess, M., & Kobie, N. (2019, March 18). *The messy, cautionary tale of how Babylon*

Canadian Medical Association (CMA). (2019, August). *Virtual care in Canada: Discussion paper*. Canadian Medical Association. https://www.cma.ca/sites/default/files/pdf/News/Virtual_Care_discussionpaper_v2EN.pdf

CBC News. (2020, March 20). Telus virtual health-care app touted by Alberta government sparks outcry from physicians. *CBC.ca*. https://www.cbc.ca/news/canada/calgary/ahs-app-covid-babylon-telus-doctor-phone-conference-reaction-1.5504317

Coleman, E. A. (2003). Falling through the cracks: Challenges and opportunities for improving transitional care for persons with continuous complex care needs. *Journal of the American Geriatrics Society, 51*(4), 549–555. https://doi.org/10.1046/j.1532-5415.2003.51185.x

Coleman, E. A., Parry, C., Chalmers, S., & Min, S. J. (2006). The care transitions intervention: Results of a randomized controlled trial. *Archives of Internal Medicine, 166*(17), 1822–1828. https://doi.org/10.1001/archinte.166.17.1822

Cooper, A., Edwards, A., Williams, H., Evans, H. P., Avery, A., Hibbert, P., Makeham, M., Sheikh, A., Donaldson, L. J., & Carson-Stevens, A. (2017). Sources of unsafe primary care for older adults: A mixed-methods analysis of patient safety incident reports. *Age and Ageing, 46*(5), 833–839. https://doi.org/10.1093/ageing/afx044

Culmer, N., Smith, T. B., Stager, C., Wright, A., Fickel, A., Tan, J., Clark, C. T., Meyer, H., & Grimm, K. (2023). Asynchronous Telemedicine: A Systematic Literature Review. *Telemedicine Reports, 4*(1), 366–386. https://doi.org/10.1089/tmr.2023.0052

Doarn, C. R., Pruitt, S., Jacobs, J., Harris, Y., Bott, D. M., Riley, W., Lamer, C., & Oliver, A. L. (2014). Federal efforts to define and advance telehealth-a work in progress. *Telemedicine and e-Health, 20*(5), 409–418. https://link.gale.com/apps/doc/A371397879/AONE?u=uniwater&sid=bookmark-AONE&xid=cffd7990

Elliott, J., Whate, A., McNeil, H., Kernoghan, A., Stolee, P., & the SHARP Group (2021). A SHARP response: Developing COVID-19 research aims in partnership with the seniors helping as research partners (SHARP) Group. *Canadian Journal on Aging = La Revue Canadienne Du Vieillissement, 40*(4), 661–668. https://doi.org/10.1017/S0714980821000453

Falk, W. (2021, June 29). *The state of virtual care in Canada as of Wave Three of the COVID-19 pandemic: An early diagnostique and policy recommendations.* Health Canada. https://www.canada.ca/content/dam/hc-sc/documents/corporate/transparency_229055456/health-agreements/bilateral-agreement-pan-canadian-virtual-care-priorities-covid-19/template-wf-report-eng.pdf

Forrest, C. B. (2003). Primary care in the United States: Primary care gatekeeping and referrals: Effective filter or failed experiment? *BMJ, 326*(7391), 692–695. https://doi.org/10.1136/bmj.326.7391.692

Forster, A. J., Murff, H. J., Peterson, J. F., Gandhi, T. K., & Bates, D. W. (2003). The incidence and severity of adverse events affecting patients after discharge from the hospital. *Annals of Internal Medicine, 138*(3), 161–167. https://doi.org/10.7326/0003-4819-138-3-200302040-00007

Friemel, T. N. (2014). The digital divide has grown old: Determinants of a digital divide among seniors. *New Media & Society, 18*(2), 313–331. https://doi.org/10.1177/1461444814538648

Ghosh, R., Lindman, D., Ratan, S., Steinmetz, V., & Center for Technology and Aging. (2014). *The new era of connected aging: A framework for understanding technologies that support older adults in aging in place.* University of California. https://healthliteracycentre.eu/wp-content/uploads/2015/11/ConnectedAgingFramework.pdf

Glazier, R. H., Green, M. E., Wu, F. C., Frymire, E., Kopp, A., & Kiran, T. (2021). Shifts in office and virtual primary care during the early COVID-19 pandemic in Ontario, Canada. *Canadian Medical Association Journal, 193*(6). https://doi.org/10.1503/cmaj.202303

Glenny, C., Stolee, P., Sheiban, L., & Jaglal, S. (2013). Communicating during care transitions for older hip fracture patients: Family caregiver and health care provider's perspectives. *International Journal of Integrated Care, 13*, e044. https://doi.org/10.5334/ijic.1076

Goldberg, D. G., Gimm, G., Burla, S. R., & Nichols, L. M. (2020). Care experiences of patients with multiple chronic conditions in a payer-based patient-centered medical home. *Population Health Management, 23*(4), 305–312. https://doi.org/10.1089/pop.2019.0189

Gonzales, A. (2016). The contemporary US digital divide: From initial access to technology maintenance. *Information, Communication & Society, 19*(2), 234–248. https://doi.org/10.1080/1369118X.2015.1050438

Hammersley, V., Donaghy, E., Parker, R., McNeilly, H., Atherton, H., Bikker, A., Campbell, J., & McKinstry, B. (2019). Comparing the content and quality of video, telephone, and face-to-face consultations: A non-randomised, quasi-experimental, exploratory study in UK primary care. *The British Journal of General Practice: THe Journal of the Royal College of General Practitioners, 69*(686), e595–e604. https://doi.org/10.3399/bjgp19X704573

Hauser-Ulrich, S., Künzli, H., Meier-Peterhans, D., & Kowatsch, T. (2020). A smartphone-based health care chatbot to promote self-management of chronic pain (SELMA): Pilot randomized controlled trial. *JMIR Mhealth and Uhealth, 8*(4), e15806. https://doi.org/10.2196/15806

Jack, B. W., Chetty, V. K., Anthony, D., Greenwald, J. L., Sanchez, G. M., Johnson, A. E., Forsythe, S. R., O'Donnell, J. K., Paasche-Orlow, M. K., Manasseh, C., Martin, S., & Culpepper, L. (2009). A reengineered hospital discharge program to decrease rehospitalization: A randomized trial. *Annals of Internal Medicine, 150*(3), 178–187. https://doi.org/10.7326/0003-4819-150-3-200902030-00007

Jimenez, G., Matcher, D., Koh, G., Tyagi, S., van der Kleij, R., Chavannes, N. H., & Car, J. (2021). Revisiting the four core functions (4Cs) of primary care: Operational definitions and complexities. *Primary Health Care Research & Development, 22*, e68. https://doi.org/10.1017/S1463423621000669

Jones, I. R. (2015). Connectivity, digital technologies and later life. In J. Twigg & W. Martin (Eds.), *Routledge handbook of cultural gerontology* (pp. 460–468). Routledge.

Judson, T. J., Subash, M., Harrison, J. D., Yeager, J., Williams, A. M., Grouse, C. K., & Byron, M. (2023). Patient Perceptions of e-Visits: Qualitative Study of Older Adults to Inform Health System Implementation. *JMIR Aging, 6*, e45641. https://doi.org/10.2196/45641

Klein, D. M., Fix, G. M., Hogan, T. P., Simon, S. R., Nazi, K. M., & Turvey, C. L. (2015). Use of the blue button online tool for sharing health information: Qualitative interviews with patients and providers. *Journal of Medical Internet Research, 17*(8), e199. PMID: 26286139. https://doi.org/10.2196/jmir.4595

Kokorelias, K. M., DasGupta, T., & Hitzig, S. L. (2021). Designing the ideal patient navigation program for older adults with complex needs: A qualitative exploration of the preferences of key informants. *Journal of Applied Gerontology: THe Official Journal of the Southern Gerontological Society, 41*(4), 1002–1010. https://doi.org/10.1177/07334648211059056

Kringos, D. S., Boerma, W. G. W., Hutchinson, A., & Saltman, R. B. (Eds.). (2015). *Building primary care in a changing Europe.* European Observatory on Health Systems and Policies.

Lafortune, C., Huson, K., Santi, S., & Stolee, P. (2015). Community-based primary health care for older adults: A qualitative study of the perceptions of clients, caregivers and health care providers. *BMC Geriatrics, 15*, 57. https://doi.org/10.1186/s12877-015-0052-x

Landi. H. (2018, April 2). *Report: Telehealth market estimated to reach $19.5B by 2025. Healthcare Informatics.* https://www.hcinnovationgroup.com/clinical-it/news/13030029/report-telehealth-market-estimated-to-reach-195b-by-2025

Macinko, J., Starfield, B., & Shi, L. (2003). The contribution of primary care systems to health outcomes within Organisation for Economic Cooperation and Development (OECD) countries, 1970–1998. *Health Services Research, 38*(3), 831–865. https://doi.org/10.1111/1475-6773.00149

Manderson, B., McMurray, J., Piraino, E., & Stolee, P. (2012). Navigation roles support chronically ill older adults through healthcare transitions: A systematic review of the literature. *Health & Social Care in the Community, 20*(2), 113–127. https://doi.org/10.1111/j.1365-2524.2011.01032.x

Martin, D. (2017). *Better now: Six big ideas to improve health for all Canadians.* Allen Lane.

Matthew-Maich, N., Harris, L., Ploeg, J., Markle-Reid, M., Valaitis, R., Ibrahim, S., Gafni, A., & Isaacs, S. (2016). Designing, implementing, and evaluating mobile health technologies for managing chronic conditions in older adults: A scoping review. *JMIR Mhealth and Uhealth, 4*(2), e29. https://doi.org/10.2196/mhealth.5127

Mercer, K., Baskerville, N., Burns, C. M., Chang, F., Giangregorio, L., Tomasson Goodwin, J., Sadat Rezai, L., & Grindrod, K. (2015). Using a collaborative research approach to develop an interdisciplinary research agenda for the study of mobile health interventions for older adults. *JMIR Mhealth and Uhealth, 3*(1), e11. https://doi.org/10.2196/mhealth.3509

Miller, S., Gilbert, S., Virani, V., & Wicks, P. (2020). Patients' utilization and perception of an artificial intelligence-based symptom assessment and advice technology in a British primary care waiting room: Exploratory pilot study. *JMIR Human Factors, 7*(3), e19713. https://doi.org/10.2196/19713

Mold, F., Hendy, J., Lai, Y.-L., & de Lusignan, S. (2019). Electronic consultation in primary care between providers and patients: Systematic review. *JMIR Medical Informatics, 7*(4), e13042. https://doi.org/10.2196/13042

National Health Service. (2019, January 8). *The NHS long term plan.* NHS. https://www.longtermplan.nhs.uk/wp-content/uploads/2019/08/nhs-long-term-plan-version-1.2.pdf

Naylor, M. D., Brooten, D. A., Campbell, R. L., Maislin, G., McCauley, K. M., & Schwartz, J. S. (2004). Transitional care of older adults hospitalized with heart failure: A randomized, controlled trial. *Journal of the American Geriatrics Society, 52*(5), 675–684. https://doi.org/10.1111/j.1532-5415.2004.52202.x

Neves, A. L., Li, E., Gupta, P. P., Fontana, G., & Darzi, A. (2021). Virtual primary care in high-income countries during the COVID-19 pandemic: Policy responses and lessons for the future. *European Journal of General Practice, 27*(1), 241–247. https://doi.org/10.1080/13814788.2021.1965120

Nilsson, E., Sverker, A., Bendtsen, P., & Eldh, A. C. (2021). A human, organization, and technology perspective on patients' experiences of a chat-based and automated medical history-taking service in primary health care: Interview study among primary care patients. *Journal of Medical Internet Research, 23*(10), e29868. https://doi.org/10.2196/29868

O'Hear, S. (2017, April 19). Ada is an AI-powered doctor app and telemedicine service. *TechCrunch+.* https://techcrunch.com/2017/04/19/ada-health/#:~:text=%E2%80%9CAda%20has%20been%20trained%20over,findings%2C%E2%80%9D%20explains%20the%20company

Pravia, C. I., & Diaz, Y. M. (2019). Primary care: Practice meets technology. *Cleveland Clinic Journal of Medicine, 86*(8), 525–528. https://doi.org/10.3949/ccjm.86a.18122

Rogers, A., Kirk, S., Gately, C., May, C. R., & Finch, T. (2011). Established users and the making of telecare work in long term condition management: Implications for health policy. *Social Science & Medicine, 72*(7), 1077–1084. https://doi.org/10.1016/j.socscimed.2011.01.031

Rusnell, C. (2021). *Telus Health ignored Alberta's privacy laws when it launched Babylon app, reports reveal.* https://www.cbc.ca/news/canada/edmonton/babylon-app-privacy-telus-health-1.6132471#:~:text=Launched%20in%20March%20in%20partnership,under%20Alberta's%20health%2Dcare%20insurance

Schmalstieg-Bahr, K., Popert, U. W., & Scherer, M. (2021). The role of general practice in complex health care systems. *Frontiers in Medicine, 8*, 680695. https://doi.org/10.3389/fmed.2021.680695

Schoen, C., Osborn, R., How, S. K. H., Doty, M. M., & Peugh, J. (2008). In chronic condition: Experiences of patients with complex health care needs, in eight countries, 2008. *Health Affairs (Project Hope), 28*(S1), w1-16. https://doi.org/10.1377/hlthaff.28.1.w1

Shaw, J., Jamieson, T., Agarwal, P., Griffin, B., Wong, I., & Bhatia, R. S. (2018). Virtual care policy recommendations for patient-centred primary care: Findings of a consensus policy dialogue using a nominal group technique. *Journal of Telemedicine and Telecare, 24*(9), 608–615. https://doi.org/10.1177/1357633X17730444

Starfield, B., Shi, L., & Macinko, J. (2005). Contribution of primary care to health systems and health. *The Milbank Quarterly, 83*(3), 457–502. https://doi.org/10.1111/j.1468-0009.2005.00409.x

Stolee, P., Elliott, J., Byrne, K., Sims-Gould, J., Tong, C., Chesworth, B., Egan, M., Ceci, C., & Forbes, D. (2019). A framework for supporting post-acute care transitions of older patients with hip fracture. *Journal of the American Medical Directors Association, 20*(4), 414–419.e1. PMID: 30852166. https://doi.org/10.1016/j.jamda.2019.01.147

Stolee, P., Elliott, J., Giguere, A. M., Mallinson, S., Rockwood, K., Sims Gould, J., Baker, R., Boscart, V., Burns, C., Byrne, K., Carson, J., Cook, R. J., Costa, A. P., Giosa, J., Grindrod, K., Hajizadeh, M., Hanson, H. M., Hastings, S., Heckman, G., … Witteman, H. (2021). Transforming primary care for older Canadians living with frailty: Mixed methods study protocol for a complex primary care intervention. *British Medical Journal Open, 11*(5), e042911. https://doi.org/10.1136/bmjopen-2020-042911

Tong, C., Kernoghan, A., Lemmon, K., Fernandes, P., Elliott, J., Sacco, V., Bodemer, S., & Stolee, P. (2022). Lessons and reflections from an extended co-design process developing an mHealth app with and for older adults: Multiphase, mixed methods study. *JMIR Aging, 5*(4), e39189. https://doi.org/10.2196/39189

Wong, A., Bhyat, R., Srivastava, S., Lomax, L. B., & Appireddy, R. (2021). Patient care during the COVID-19 pandemic: Use of virtual care. *Journal of Medical Internet Research, 23*(1), e20621. https://doi.org/10.2196/20621

World Health Organization & Global Observatory for eHealth. (2011). *mHealth: New horizons for health through mobile technologies. Second global survey on eHealth.* WHO Press. https://apps.who.int/iris/handle/10665/44607

World Health Organization & the United Nations Children's Fund. (2020). *Operational framework for primary healthcare: Transforming vision into action.* World Health Organization, Geneva. https://www.who.int/publications/i/item/9789240017832

Young, H. M., & Nesbitt, T. S. (2017). Increasing the capacity of primary care through enabling technology. *Journal of General Internal Medicine, 32*, 398–403. https://doi.org/10.1007/s11606-016-3952-3

Part III
Part Three

8 Electronic Health Records

The Challenge

Electronic health records (EHRs) offer many potential benefits for the care of older persons, including applications for decision support, care coordination, quality improvement, and research. The benefits of EHRs for older adults have not been fully realized due to slow and uneven adoption, lack of interoperability, and inadequate consideration of older adults' needs and preferences in their design (Fig. 8.1).

Persona and Scenario—Dominic

Dominic is an 84-year old man who lives alone; his wife passed away six years ago. He uses a cane, is a bit forgetful, and needs some help from his son Luca and daughter-in-law Amalia with housework, shopping and medical appointments. He has a few health problems, including hypertension, osteoporosis and arthritis. Recently he began taking medication for an enlarged prostate. He finds this makes him feel dizzy when he stands up, but Dominic feels this effect will go away over time. Unfortunately, he got up quickly to answer the phone and the resulting dizziness caused him to fall heavily and break his hip. Dominic manages to reach the phone and call the emergency number. An ambulance arrives to take him to hospital where he waits on a stretcher for 24 h before going into surgery. While waiting on a stretcher, Dominic is able to provide his name and his son's name but cannot remember all of the medications he is on or the name of his family doctor. After surgery, he receives visits from the surgeon and several nurses, but it is unclear how long he will be in hospital or what will happen next. A week later he is told that he will be transferred that day to a rehabilitation hospital. He is surprised because he thought he would be going home and wonders how his son and daughter-in-law will

D. Juzwishin et al., *AgeTech Innovations in Healthcare for Older Adults*, Synthesis Lectures on Technology and Health, https://doi.org/10.1007/978-3-031-86885-6_8

Fig. 8.1 Accessing information in online patient portal. *Credit* Shutterstock: Halfpoin, https://www.shutterstock.com/image-photo/senior-woman-learning-work-tablet-technology-2461284687

know where he is. At the rehabilitation hospital, his therapy is delayed while they wait for his medical records to arrive. Two weeks later he is discharged home to continue his therapy, but information on his health conditions and therapy recommendations are not communicated with the home care program. A physiotherapist comes to the home and conducts an assessment to develop a new exercise program. Dominic has trouble remembering the exercises. One day Dominic decides to take a bath without help, slips and falls in the tub. He fractures his wrist trying to break his fall. Following surgery, Luca and Amalia begin to discuss long-term care admission with Dominic.

Electronic Health Records

Electronic health records (EHRs) are an information technology that can be applied in all sectors of the health care system. An EHR has been defined as "a digital version of a patient's chart. EHRs are real-time, patient-centered records that make information available instantly and securely to authorized users" (Office of the National Coordinator for Health Information Technology, 2019, para. 1).

EHR Use for Clinical Practice and Patient Safety

Electronic health records (EHRs) can play an important role in care delivery and care coordination, and can support linkages, communication, and referrals between health services and facilities (WHO, 2018). Marchibroda (2008) has described the benefits of health information technology in chronic care, including timely and secured access to patient information. Falconer and colleagues (2018) found that EHRs were commonly used for care coordination for persons with mental health issues, including older adults, and particularly in primary care settings. Advantages include increased shared access to patient information, and improved capacity for scheduling and screening. A systematic review of the role of digital health technologies in primary care found that EHRs were the most commonly used technology in primary care to improve continuity and comprehensiveness of care, and that their use reduced the need for specialist consultation care (Jimenez et al., 2021). A United States-based study involving interviews of informants from 27 primary care practices found that EHRs facilitated communication and teamwork within health care teams in several ways, including through instant messaging and delegation of tasks, and templates to support symptom specific data collection (O'Malley et al., 2015).

Online platforms, such as Caredove, are being used to support primary care referrals to home and community care agencies (Stolee et al., 2021).

Caredove

Caredove (www.caredove.ca) is an online database of health care services and community resources. Referrals to these resources can be made directly through the platform, which also allows for follow-up to determine whether patients are receiving the service. Stolee et al. (2021) studied the implementation of Caredove in primary care settings in Canada, with identification of appropriate resources guided by a screening algorithm.

Electronic health information systems and information technology have important applications for clinical decision-making and patient safety (Bates & Gawande, 2003). These include their use for computerized provider order entry (CPOE) for medications and lab tests (Dhamanti et al., 2021); medication alerts and reconciliation tools (Iankowitz et al., 2012; Tsilimingras & Bates, 2008); evidence alerts and clinical decision support systems (Alagiakrishnan et al., 2019; Bates & Gawande, 2003); and database-generated discharge summaries (Hesselink et al., 2012). To enhance the value of EHRs for use in geriatric care, a recent survey of health care providers identified a need to enhance geriatrics-specific documentation of cognition, functional status and goals of care (Southerland et al., 2023).

The use of EHRs for quality improvement has been described as a key building block of high-performing primary care (Bodenheimer et al., 2014). EHRs can incorporate health care quality measures of benefit to older persons, such as for depression care (Campbell et al., 2021), and can support prevention, such as initiatives to prevent falls (Sarmiento & Lee, 2017).

The above applications suggest significant potential for EHRs to improve patient outcomes, however the evidence in some sectors is ambiguous. Dendere and colleagues (2021) found that the introduction of EHRs into residential care facilities had both potential positive and negative effects and recommended further investigation of their impact on outcomes.

EHRs and other eHealth initiatives have been found to support patient transitions between settings, including monitoring and follow-up (Fjellså et al., 2022). A European study found that EHR systems with patient registration databases and tools for communication between care provides can support integrated care for older patients with multimorbidity (Melchiorre et al., 2018).

EHRs have potential applications in research that can yield new knowledge that can benefit the quality of care and quality of life of older persons. Andrew and colleagues (2023) have described the development of an EHR derived data platform that enabled research into healthy aging. Bowman and colleagues (2020) used primary care EHR data to identify factors predictive of delirium in the community.

Online Patient Portals

Online patient portals can provide patients with access to their EHRs. A review by Lyles et al. (2020) found that while many health care systems offer portals, relatively few patients access these. Wider adoption will require one-on-one patient training, accommodation of patients with limited health or digital literacy, and support to clinicians to incorporate patient use of portals in their practice (Lyles et al., 2020).

Piraino et al. (2017) explored the potential of linking service providers to an online family caregiver network. The network was seen as an opportunity to fill communication gaps, but barriers were identified related to time, privacy and policy restrictions. Some portals, such as "Elizz" developed by SE Health (elizz.com), provide information, skills training and support for caregivers. Wolff et al. (2022) have explored the challenges and opportunities of engaging the care partners of older adults in patient portals and emphasized the need for appropriate privacy and security protocols.

A Finnish study (Huvila et al., 2022) found age-related variation in the use of patient-accessible EHRs. Huvila et al. (2022) found that younger persons expressed greater general interest but that older persons were more likely to use their access to their EHR to understand their health conditions and to become more involved in their health care. A literature review by Wildenbos et al. (2017) found that patient portals were more likely to be accepted by older adults if their use was voluntary and if they thought their use would be beneficial. Barriers to use included limited health literacy, limited technology access and privacy concerns. The use of portals by older patients would be enhanced by greater consideration of the needs and preferences of older adults in their design (Wang & Zhao, 2023; Zoorob et al., 2022).

Steele Gray and colleagues (2021) are developing an electronic patient-reported outcome tool, the ePRO mobile app and portal system, designed to support goal-oriented care of older adults with complex health needs in primary care practices. Although a promising model, its implementation in a complex setting has faced significant challenges including the level of readiness of providers and patients.

Embedded Risk and Frailty Screening

In the chapter of this book on long term residential care, we described a role for standardized assessment systems, such as the interRAI assessment tools, to support clinical decision-making and quality improvement. There has been less use of such systems in primary care settings, although there has been increasing interest in risk and frailty screening tools in primary care, and in embedding these tools within electronic information systems. The interRAI consortium have developed a screening tool that supports the Assessment Urgency Algorithm (AUA) to classify patients into risk categories (Hirdes, 2006). This tool has been used in emergency department settings (Gray et al., 2013); its use in primary care is currently being explored (Stolee et al., 2021). Whate and colleagues (2021) have studied the use of this tool for risk screening of older adults by paramedics and found that the AUA score predicted multiple adverse outcomes and may be useful in earlier identification of older adults needing additional intervention. Of particular relevance to the use of EHRs, Clegg and colleagues (2016) have developed an electronic version of the Frailty Index (FI) developed by Rockwood and Mitnitski (2007). The FI is based on a model of cumulative deficits and is calculated as an individual's number of health conditions, symptoms and disabilities as a proportion of a large number of measured deficits. While extensively validated (e.g., Song et al., 2010), the FI would be unwieldy for use in routine clinical practice. The eFI however, is automatically populated from routinely collected data already contained in EHRs.

Potential Benefit of EHRs Unrealized

Many of the potential benefits of EHRs for care coordination and other applications remain unrealized due to the lack of interoperability of many EHR systems across settings and sectors (Birke et al., 2020; McMurray et al., 2013; Persaud, 2019) and the sustained use of paper-based record systems (McMurray et al., 2013). McMurray and colleagues (2013) conducted an ethnographic investigation of electronic information system use related to care transitions of older patients. They found that the predominant mode for communicating health information during care transitions was paper-based even when EHRs are present; they also found that duplicate (electronic and paper) documentation was common. Many organizations, such as LTC homes, did not use EHRs, thus information associated with the transfer of older persons to the home was done using paper documentation sent via mail or fax or verbal communication (McMurray et al., 2013). In the absence of electronic systems, clinicians may rely on time-consuming manual methods for searching and sifting through written documentation, and informal systems for communicating with other care providers and settings (Stolee et al., 2019). Inconsistent and inadequate communication can result in complex older patients "falling through the cracks" (Coleman, 2003) when transitioning between care settings.

Canadian Physician Argues for a National Electronic Health Record for Primary Care

"The good news is that most primary care providers now use an electronic health record. The bad news is that there was no coordination in selecting them … The past decade has seen many reports written about "interoperability," setting standards so data can be shared between systems, but repeating the buzzword did not bring us actual interoperability. The large number of attempted solutions has become an additional problem. So much for a single health record for each patient.

Rather than adding another layer to an overly complicated system, we should rebuild it from the ground up, starting with primary care records. We should select a single system and make it the national primary care electronic health record. If all primary care providers used it, the billions already spent on electronic health records themselves—or trying to get the different ones to talk to each other—could be spent on improving the one everyone used" (Persaud, 2019, p. E28).

Factors Affecting EHR Adoption

Adoption of EHRs in primary care and other settings has been slow (O'Donnell et al., 2018; Palabindala et al., 2016), and particularly slow for the use of more advanced EHR features (Rahal et al., 2021). Jiang et al. (2023) found that adoption of EHRs in US hospitals is increasing, but that significant disparities remain. A systematic review found that EHR implementation in low-income countries was at an early stage (Woldemariam & Jimma, 2023). The 2021 OECD survey reported increases in the use of EHRs but persistent issues with system fragmentation, provider resistance, technical barriers and legal issues (Slawomirski et al., 2023).

Wider adoption of electronic health information systems is affected by a range of factors. Ludwick and Doucette (2009) conducted a systematic literature review of the electronic information system adoption experiences in seven countries. They found that implementation was affected by design quality and functionality, project management and procurement, and prior experiences; common concerns included "privacy, patient safety, provider and patient relations, staff anxiety, time factors, quality of care, finances, efficiency, and liability" (Ludwick & Doucette, 2009, p. 22).

Promising Initiatives

Stimulated by the COVID-19 pandemic and Sustainable Development Goal (SDG) 3.8 to achieve universal health coverage, many countries are undertaking initiatives to build a backbone of an EHR to support health care delivery for their citizens. To facilitate interoperability, common standards are being adopted supported by strong data governance and regulatory frameworks. The Netherland's health system demonstrates how efforts through regional initiatives and projects led and supported by a national digital health strategy can accomplish greater alignment of policies and procedures to ensure privacy and promote data security (Organization for Economic Co-operation & Development, 2022).

In Canada, prompted by the COVID-19 pandemic, the national government released the Pan-Canadian Health Data Strategy identifying the need to strengthen the health data foundations by clarifying accountability and health data governance so that Canadians can benefit from the advantages of interoperable EHRs (Public Health Agency of Canada, 2022). A Pan-Canadian Health Data Charter (Health Canada, 2023) has been developed that establishes 10 principles built around the commitment to citizen-centric health information architecture to ensure that health data follow the individual across all types and levels of care.

Digitization of health data brings many promises but also brings significant risk of security and confidentiality breaches that require that legislators, regulators and policy makers are diligent in developing robust systems. Guidance for how to address the challenges in the future are provided in a dedicated issue of a recent health care management periodical (Landre, 2024).

How Could These Technologies Help Dominic?

Dominic could benefit from a well-developed and integrated system of EHRs.

By the time Dominic arrives in the emergency department, information collected by the paramedics has been transferred to the hospital. Notifications are sent to his primary care physician and his son Luca. The primary care physician's own electronic health record provides alerts on Dominic's frailty status and information on his medications and health conditions. Following surgery, information on his condition and plan of care is available on an online patient portal. Dominic has given Luca permission to access the portal so that he can advise and support his father. The integrated EHR systems and patient portal support timely communication of assessment information and care plans to the family, the rehabilitation hospital, the primary care physician, and the home care program. With support from his health care providers and family, Dominic's rehabilitation is successful, and he continues living in his own home.

Seminal Demonstration Project

In many health care settings, patient health information is gathered during clinical assessments and laboratory investigations and recorded in EHRs. These data can be used to generate indicators of frailty or risk, such as the eFI or the AUA, which can prompt more detailed assessments or referrals to specialized services. A limitation of these data is that they are collected intermittently and do not permit continuous monitoring of an older patient's health status or physical activity. Recent work has demonstrated an association of data collected using wearable devices with data obtained through clinical assessments. Kim et al. (2021) found an association between frailty and data collected using a consumer-grade wearable device. Yang and colleagues (2019) undertook a small-scale observational study in which risk of falls of community-dwelling older persons (n = 40 participants, aged 65 to 93) was estimated using (1) physical activity data gathered using a wrist-worn electronic device; (2) data gathered using the interRAI home care assessment, the RAI-HC, which is collected at 90-day intervals; and (3) a combination of the wearable and RAI-HC data. The study found that the wearable and assessment data were both able to predict fall risk, but that the best classification performance was achieved with a combination of both sources of data.

We propose that a demonstration project be undertaken building on the promising results of the studies by Kim et al. (2021) and Yang et al. (2019). This will require systems to integrate monitoring data, collected in real time, into the electronic record, and to combine the two sources of data using algorithms to predict risk of adverse events such as falls or to identify patients who require urgent assessment.

Diverse groups of patients, family caregivers and health care practitioners should be engaged in planning the project, including selection of health data to be monitored, wearable devices to be used, the risk (e.g., of falls) to be predicted, and the criteria for success. The project should address technical support and training requirements for patients and practitioners. Sustainability should be considered, including costs of the technologies and system integration, and how these can be funded on a long term basis.

References

Alagiakrishnan, K., Ballermann, M., Rolfson, D., Mohindra, K., Sadowski, C. A., Ausford, A., Romney, J., & Hayward, R. S. (2019). Utilization of computerized clinical decision support for potentially inappropriate medications. *Clinical Interventions in Aging, 14*, 753–762. https://doi.org/10.2147/CIA.S192927

Andrew, N. E., Beare, R., Ravipati, T., Parker, E., Snowdon, D., Naude, K., & Srikanth, V. (2023). Developing a linked electronic health record derived data platform to support research into healthy ageing. *International Journal of Population Data Science, 8*(1), 2129. https://doi.org/10.23889/ijpds.v8i1.2129

Bates, D. W., & Gawande, A. A. (2003). Improving safety with information technology. *New England Journal of Medicine, 348*, 2526–4. https://doi.org/10.1056/NEJMsa020847

Birke, H., Jacobsen, R., Jønsson, A. B., Guassora, A., Walther, M., Saxild, T., Laursen, J. T., Vall-Lamora, M., & Frølich, A. (2020). A complex intervention for multimorbidity in primary care: A feasibility study. *Journal of Comorbidity, 10*, 2235042X20935312. https://doi.org/10.1177/2235042X20935312

Bodenheimer, T., Ghorob, A., Willard-Grace, R., & Grumbach, K. (2014). The 10 building blocks of high-performing primary care. *Annals of Family Medicine, 12*(2), 166–1. https://doi.org/10.1370/afm.1616

Bowman, K., Jones, L., Masoli, J., Mujica-Mota, R., Strain, D., Butchart, J., Valderas, J. M., Fortinsky, R. H., Melzer, D., & Delgado, J. (2020). Predicting incident delirium diagnoses using data from primary-care electronic health records. *Age and Ageing, 49*(3), 374–381. https://doi.org/10.1093/ageing/afaa006

Campbell, C. M., Murphy, D. R., Taffet, G. E., Major, A. B., Ritchie, C. S., Leff, B., & Naik, A. D. (2021). Implementing health care quality measures in electronic health records: A conceptual model. *Journal of the American Geriatrics Society, 69*(4), 1079–1085. https://doi.org/10.1111/jgs.17033

Clegg, A., Bates, C., Young, J., Ryan, R., Nichols, L., Ann Teale, E., Mohammed, M. A., Parry, J., & Marshall, T. (2016). Development and validation of an electronic frailty index using routine primary care electronic health record data. *Age and Ageing, 45*(3), 353–60. https://doi.org/10.1093/ageing/afw039

Coleman, E. A. (2003). Falling through the cracks: Challenges and opportunities for improving transitional care for persons with continuous complex care needs. *Journal of the American Geriatrics Society, 51*(4), 549–55. https://doi.org/10.1046/j.1532-5415.2003.51185.x

Dendere, R., Samadbeik, M., & Janda, M. (2021). The impact on health outcomes of implementing electronic health records to support the care of older people in residential aged care: A scoping review. *International Journal of Medical Informatics, 151*, 104471. https://doi.org/10.1016/j.ijmedinf.2021.104471

Dhamanti, I., Kurniawati, E., Zairina, E., Nurhaida, I., & Salsabila, S. (2021). Implementation of computerized physician order entry in primary care: A scoping review. (2021). *Journal of Multidisciplinary Healthcare, 14*, 3441–51. https://doi.org/10.2147/JMDH.S344781

Falconer, E., Kho, D., & Docherty, J. P. (2018). Use of technology for care coordination initiatives for patients with mental health issues: A systematic literature review. *Neuropsychiatric Disease and Treatment, 14*, 2337–49. https://doi.org/10.2147/NDT.S172810

Fjellså, H. M. H., Husebø, A. M. L., & Storm, M. (2022). EHealth in care coordination for older adults living at home: Scoping review. *Journal of Medical Internet Research, 24*(10), e39584. https://doi.org/10.2196/39584

Gray, L. C., Peel, N. M., Costa, A. P., Burkett, E., Dey, A. B., Jonsson, P. V., Lakhan, P., Ljunggren, G., Sjostrand, F., Swoboda, W., Wellens, N., & Hirdes, J. (2013). Profiles of older patients in the emergency department: Findings from the interRAI multinational emergency department study. *Annals of Emergency Medicine, 62*(5), 467–4. https://doi.org/10.1016/j.annemergmed.2013.05.008

Health Canada. (2023). Pan-Canadian health data charter. Health Canada. https://www.canada.ca/en/health-canada/corporate/transparency/health-agreements/shared-health-priorities/working-together-bilateral-agreements/pan-canadian-data-charter.html

Hesselink, G., Schoonhoven, L., Barach, P., Spijker, A., Gademan, P., Kalkman, C., Liefers, J., Vernooij-Dassen, M., & Wollersheim, H. (2012). Improving patient handovers from hospital to primary care: A systematic review. *Annals of Internal Medicine, 157*(6), 417–28. https://doi.org/10.7326/0003-4819-157-6-201209180-00006. PMID: 22986379

Hirdes, J. P. (2006). Addressing the health needs of frail elderly people: Ontario's experience with an integrated health information system. *Age and Ageing, 35*(4), 329–1. https://doi.org/10.1093/ageing/afl036

Huvila, I., Cajander, Å., Moll, J., Enwald, H., Eriksson-Backa, K., & Rexhepi, H. (2022). Technological and informational frames: Explaining age-related variation in the use of patient accessible electronic health records as technology and information. *Information Technology and People, 35*(8), 1–22. https://doi.org/10.1108/ITP-08-2020-0566

Iankowitz, N., Dowden, M., Palomino, S., Uzokwe, H., & Worral, P. (2012). The effectiveness of computer system tools on potentially inappropriate medications ordered at discharge for adults older than 65 years of age: A systematic review. *The Joanna Briggs Institute Library of Systematic Reviews, 10*(13), 798–31. https://doi.org/10.11124/jbisrir-2012-68

Jiang, J., Qi, K., Bai, G., & Schulman, K. (2023). Pre-pandemic assessment: A decade of progress in electronic health record adoption among U.S. hospitals, *Health Affairs Scholar, 1*(5). https://doi.org/10.1093/haschl/qxad056

Jimenez, G., Matchar, D., Koh, C., van der Kleij, R., Chavannes, N. H., & Car, J. (2021). The role of health technologies in multicomponent primary care interventions: Systematic review. *Journal of Medical Internet Research, 23*(1), e20195. https://doi.org/10.2196/20195

Kim, B., Hunt, M., Muscedere, J., Maslove, D. M., & Lee, J. (2021). Using consumer-grade physical activity trackers to measure frailty transitions in older critical care survivors: Exploratory observational study. *JMIR Aging, 4*(1), e19859. https://doi.org/10.2196/19859

Landre, J. (2024). A message from the guest editor. *Healthcare Management Forum, 37*(1), 3–4. https://doi.org/10.1177/08404704231198352

Ludwick, D. A., & Doucette, J. (2009). Adopting electronic medical records in primary care: Lessons learned from health information systems implementation experience in seven countries. *International Journal of Medical Informatics, 78*(1), 22–31. https://doi.org/10.1016/j.ijmedinf.2008.06.005

Lyles, C. R., Nelson, E. C., Frampton, S., Dykes, P. C., Cemballi, A. G., & Sarkar, U. (2020). Using electronic health record portals to improve patient engagement: Research priorities and best practices. *Annals of Internal Medicine, 172*(11 Suppl), S123-9. https://doi.org/10.7326/M19-0876

Marchibroda, J. M. (2008). The impact of health information technology on collaborative chronic care management. *Journal of Managed Care Pharmacy: JMCP, 14*(2 Suppl), S3-11.

McMurray, J., Hicks, E., Johnson, H., Elliott, J., Byrne, K., & Stolee, P. (2013). 'Trying to find information is like hating yourself every day: The collision of electronic information systems in transition with patients in transition. *Health Informatics Journal, 19*(3), 218–32. https://doi.org/10.1177/1460458212467547

Melchiorre, M. G., Papa, R., Rijken, M., van Ginneken, E., Hujala, A., & Barbabella, F. (2018). EHealth in integrated care programs for people with multimorbidity in Europe: Insights from the ICARE4EU project. *Health Policy (Amsterdam, Netherlands), 122*(1), 53–63. https://doi.org/10.1016/j.healthpol.2017.08.006

O'Donnell, A., Kaner, E., Shaw, C., & Haighton, C. (2018). Primary care physicians attitudes to the adoption of electronic medical records: A systematic review and evidence synthesis using the clinical adoption framework. *BMC Medical Informatics and Decision Making, 18*, Article no. 101. https://doi.org/10.1186/s12911-018-0703-x

O'Malley, A. S., Draper, K., Gourevitch, R., Cross, D. A., & Scholle, S. H. (2015). Electronic health records and support for primary care teamwork. *Journal of the American Medical Informatics Association (JAMIA), 22*(2), 426–4. https://doi.org/10.1093/jamia/ocu029

Office of the National Coordinator for Health Information Technology. (2019). [Website]. Office of the national coordinator for health information technology. https://www.healthit.gov/faq/what-electronic-health-record-ehr

Organization for Economic Co-operation and Development. (2022). *Towards an integrated health information system in the Netherlands.* OECD Publishing. https://doi.org/10.1787/44f43851-en

Palabindala, V., Pamarthy, A., & Jonnalagadda, N. R. (2016). Adoption of electronic health records and barriers. *Journal of Community Hospital Internal Medicine Perspectives, 6*(5), 32643. https://doi.org/10.3402/jchimp.v6.32643

Persaud N. (2019). A national electronic health record for primary care. *CMAJ: Canadian Medical Association Journal / journal de l'Association medicale canadienne, 191*(2), E28–9. https://doi.org/10.1503/cmaj.181647

Piraino, E., Byrne, K., Heckman, G. A., & Stolee, P. (2017). Caring in the information age: Personal online networks to improve caregiver support. *Canadian Geriatrics Journal: CGJ, 20*(2), 85–3. https://doi.org/10.5770/cgj.20.271

Public Health Agency of Canada (PHAC). (2022). *Moving forward on a Pan-Canadian health data strategy.* PHAC. https://www.canada.ca/en/public-health/programs/pan-canadian-health-data-strategy.html

Rahal, R. M., Mercer, J., Kuziemsky, C., & Yaya, S. (2021). Factors affecting the mature use of electronic medical records by primary care physicians: A systematic review. *BMC Medical Informatics and Decision Making, 21*(1), 67. https://doi.org/10.1186/s12911-021-01434-9

Rockwood, K., & Mitnitski, A. (2007). Frailty in relation to the accumulation of deficits. *The Journals of Gerontology. Series A, Biological Sciences and Medical Sciences, 62*(7), 722–7. https://doi.org/10.1093/gerona/62.7.722

Sarmiento, K., & Lee, R. (2017). STEADI: CDC's approach to make older adult fall prevention part of every primary care practice. *Journal of Safety Research, 63*, 105–109. https://doi.org/10.1016/j.jsr.2017.08.003

Slawomirski, L., Lindner, L., de Bienassis, K., et al. (2023). *Progress on implementing and using electronic record systems: Developments in OECD countries as of 2021. OECD health working papers No. 160.* Organization for Economic Co-operation and Development.

Song, X., Mitnitski, A., & Rockwood, K. (2010). Prevalence and 10-year outcomes of frailty in older adults in relation to deficit accumulation. *Journal of the American Geriatrics Society, 58*(4), 681–7. https://doi.org/10.1111/j.1532-5415.2010.02764.x

Southerland, L. T., Presley, C. J., Hunold, K. M., Caterino, J. M., Collins, C. E., & Walker, D. M. (2023). Barriers to and recommendations for integrating the age-friendly 4-Ms framework into electronic health records. *Journal of the American Geriatrics Society, 71*(5), 1573–1579. https://doi.org/10.1111/jgs.18156

Steele Gray, C., Chau, E., Tahsin, F., Harvey, S., Loganathan, M., McKinstry, B., Mercer, S. W., Nie, J. X., Palen, T. E., Ramsay, T., Thavorn, K., Upshur, R., & Wodchis, W. P. (2021). Assessing the implementation and effectiveness of the electronic patient-reported outcome tool for older adults with complex care needs: Mixed methods study. *Journal of Medical Internet Research, 23*(12), e29071. https://doi.org/10.2196/29071

Stolee, P., Elliott, J., Byrne, K., Sims-Gould, J., Tong, C., Chesworth, B., Egan, M., Ceci, C., & Forbes, D. (2019). A framework for supporting post-acute care transitions of older patients with hip fracture. *Journal of the American Medical Directors Association, 20*(4), 414–9.e1. https://doi.org/10.1016/j.jamda.2019.01.147. PMID: 30852166

Stolee, P., Elliott, J., Giguere, A. M., Mallinson, S., Rockwood, K., Sims Gould, J., Baker, R., Boscart, V., Burns, C., Byrne, K., Carson, J., Cook, R. J., Costa, A. P., Giosa, J., Grindrod, K., Hajizadeh, M., Hanson, H. M., Hastings, S., Heckman, G., … Witteman, H. (2021). Transforming primary care for older Canadians living with frailty: Mixed methods study protocol for a complex primary care intervention. *British Medical Journal Open, 11*(5), e042911. https://doi.org/10.1136/bmjopen-2020-042911

Tsilimingras, D., & Bates, D. W. (2008). Addressing postdischarge adverse events: A neglected area. *The Joint Commission Journal on Quality and Patient Safety, 34*(2), 85–97. https://doi.org/10.1016/s1553-7250(08)34011-2

Wang, X., & Zhao, Y. C. (2023). Understanding older adults intention to use patient-accessible electronic health records: Based on the affordance lens. *Frontiers in Public Health, 10*, 1075204. https://doi.org/10.3389/fpubh.2022.1075204

Whate, A., Elliott, J., Carter, D., & Stolee, P. (2021). Performance of the interRAI ED screener for risk-screening in older adults accessing paramedic services. *Canadian Geriatrics Journal: CGJ, 24*(1), 8–13. https://doi.org/10.5770/cgj.24.451

Wildenbos, G. A., Peute, L., & Jaspers, M. (2017). Facilitators and barriers of electronic health record patient portal adoption by older adults: A literature study. *Studies in Health Technology and Informatics, 235*, 308–312.

Woldemariam, M. T., & Jimma, W. (2023). Adoption of electronic health record systems to enhance the quality of healthcare in low-income countries: A systematic review. *BMJ Health and Care Informatics, 30*(1), e100704. https://doi.org/10.1136/bmjhci-2022-100704

Wolff, J. L., Dukhanin, V., Burgdorf, J. G., & DesRoches, C. M. (2022). Shared access to patient portals for older adults: Implications for privacy and digital health equity. *JMIR Aging, 5*(2), e34628. https://doi.org/10.2196/34628

World Health Organization and United Nations Children's Fund. (2018). *A vision for primary health care in the 21st century: Towards universal health coverage and the sustainable development goals.* World Health Organization. https://apps.who.int/iris/handle/10665/328065

Yang, Y., Hirdes, J. P., Dubin, J. A., & Lee, J. (2019). Fall risk classification in community-dwelling older adults using a smart wrist-worn device and the resident assessment instrument-home care: Prospective observational study. *JMIR Aging, 2*(1), e12153. https://doi.org/10.2196/12153

Zoorob, D., Hasbini, Y., Chen, K., Wangia-Anderson, V., Moussa, H., Miller, B., & Brobst, D. (2022). Ageism in healthcare technology: The older patients aspirations for improved online accessibility. *JAMIA Open, 5*(3), ooac061. https://doi.org/10.1093/jamiaopen/ooac061

9 Aging and Falling Through the Cracks of Health Care Delivery

The Challenge

A democratic civil society, entering a social contract with its citizens confers a right to health, safety, and security that the state is obligated to provide. Citizens on the other hand have a responsibility to contribute to the healthy functioning of a civil society. The respective rights and responsibilities of governments, organizations, and citizens are prescribed and exercised through legislation, regulations, policies and practices that encourage a balance between individual agency and communitarian behaviour (Fig. 9.1).

In this chapter we focus on the unmet health care needs of the older adults in society that do not fit into the conventional provision of care identified in the previous chapters. They may be identified as elder orphans, homeless in some cases or older persons without advocates. They are often socially and physically isolated and living without formal or familial support. Elder orphans may arise because they may be estranged from or have lost a family member, employment or friends. This older adult may not have a family practitioner, social worker or nurse practitioner assigned to them who would serve as the point of contact for access to and coordination of heath care services. This may be because of human resource constraints in program delivery or other factors such as isolation, geography and/or lack of transportation.

Older adults in this interstitial space are often disconnected from access to primary care and subsequent referral, triaging and transition to other types and levels of necessary care and support. The absence of support to these persons may be exacerbated by the absence of targeted program delivery, coordination and or cooperation among policymakers, funders, providers leading to further marginalization of the at risk older adults.

The digital divide between older adults generally and older adults living in poverty specifically means that many may not receive the benefit of much needed services

D. Juzwishin et al., *AgeTech Innovations in Healthcare for Older Adults*, Synthesis Lectures on Technology and Health, https://doi.org/10.1007/978-3-031-86885-6_9

Fig. 9.1 Support when it's needed most. *Credit* Shutterstock: Streamlight Studio, https://www.shutterstock.com/image-photo/portrait-asian-man-sad-mood-his-1762828835

(Alkureishi et al., 2021). This chapter will identify and argue for health care and social support reforms coupled with the optimal use of AgeTech to provide a means to reach and better serve these at-risk orphan subpopulations.

Our foci are the aging adult living in poverty who may fall through the cracks of the health care delivery systems. For this book age 50 has been used as an age cut-off for "older" homeless persons (McDonald et al., 2007).

Persona and Scenario—Francisco

Francisco is 51 years old and is homeless in Sydney. He is on income assistance but has challenges completing a monthly report to qualify for funding. Francisco, formerly a schoolteacher, had his mental illness treated successfully until his spouse died in a surfing accident. His connection to his social worker is sporadic and unpredictable. He starts his day at 5:30 a.m. with the unending tasks of securing his safety, food, shelter and companionship. None are assured. In many countries the number of homeless with a high proportion of older people if they live that long is astounding). After awaking in a cardboard and tarp shelter in the river valley Francisco ascends to street level of the city where he begins to look for warmth and hygiene before the city stirs to life. His first stop is the newly built public toilets where he can clean himself up in private and start his day. Francisco has held down menial jobs in the past, but his mental health state is such that he very quickly alienates those he works with and is currently unemployed. He continues to search for jobs stopping into businesses or going to the public library to search for jobs on the Internet. The library is a warm place where bathrooms are accessible, and one can rest and read. During the cooler and wetter season Francisco will use public transportation to travel from one spot to another to stay warm and dry. Social income assistance is available to Francisco, but he does not comply with or is unable to interact with the system in place to maintain his registration. Lunch for Francisco is earned by panhandling for money on the street. For dinner he attends the soup kitchen in the downtown core. A homeless shelter is not in the cards for Francisco because his drug addiction causes him to behave erratically and be unable to meet the criteria of the shelter. How can AgeTech support Francisco?

We have chosen a highly marginalized persona and scenario that may or may not be the most representative of the aging adult living in poverty in your community. We do this deliberately, if health reforms coupled with AgeTech can pass the test of effectively helping Francisco address the issues of access for persons experiencing homelessness it ought to be easier to introduce it in other settings that have more supports surrounding them. In many cases the marginalization of persons like Francisco is not their "fault" but reflects a lifelong history that they have had to navigate and negotiate. Secondly, persons like Francisco often become the target of a moral judgement which diminishes their self-worth. In a civil society everyone deserves the dignity and respect where they have access to appropriate care.

Finally, from a pragmatic perspective the marginalized in society may make up a very small proportion of the population yet their health and medical needs if not addressed on a timely basis can contribute to a very high cost to the health care system. When a society emphasizes responses to health problems through the medical model of interventions rather than disease prevention or health promotion it pays a high price downstream both in terms of costs of care and poor quality of life for the older adult.

Who Are the Older Adults Living in Poverty?

What is meant by the term aging adult living in poverty? Hall and colleagues proposed marginalization as a nursing theory in 1994. Initially seven properties were identified as "properties of marginalization—intermediacy, differentiation, power, secrecy, reflectiveness, voice and liminality" (Hall et al., 1994, p. 23). Over the next 25 years marginalization theory expanded taking on a global perspective and additional properties were added. Research teams used marginalization to study patient subpopulations such as homeless persons and substance abusers. In 2019 an integrative review Baah et al. used intersectionality theory to identify and review definitions of marginalization (Baah et al.,

2019). They found that three themes depicted marginalization as a phenomenon through which population groups experience multiple social determinants which limit their access to the health care delivery systems. The three themes are: creation of margins; living between cultures; and creation of vulnerabilities (Baah et al., 2019). The creation and maintenance of margins or boundaries serve as barriers between individuals and the environment or the institutions. In our case the health care delivery system creates a partition between the dominant and peripheralized group resulting in a life "between cultures" (Baah et al., 2019, p. 4):

> Not only are these individuals faced with the stark reality of thriving in a low-resourced community, but they also experience both a physical and a psychological struggle for survival along the intersections of multiple vulnerabilities. For instance, vulnerable populations who exist at these intersections may struggle with a constant need to choose between routine health checks and paying monthly rent or refilling medications and nutritious foods. These struggles make it difficult for vulnerable groups to adjust to the mainstream society.

The cumulative effect of living in poverty and living between cultures creates states of vulnerability in the subpopulation. An optimally functioning health care system must be prepared to find every means at its disposal to reach out to older adults experiencing poverty. AgeTech along with needed health care delivery reforms may help respond to this challenge.

Where and What Is the Interstitial Space?

The Merriam-Webster (2023) online dictionary defines *interstitial* as "occurring in or being an interval or intervening space or segment." In the clinical sense it is "pertaining to being between things." In the societal context we mean those safe and secure spaces that those people experiencing poverty may not have a welcoming access to. Health care delivery systems are attempting to reach out to people experiencing poverty and ensure that they receive the same services and benefits that most of the citizens have access to. Traditionally, clients are expected to come to the providers to access services, however, when the characteristics or conditions of your marginalization prevent you from entering the institution or you perceive that you are unwelcome in that institutional setting you are not likely to receive its benefits.

Coupled with appropriate and necessary reforms AgeTech may provide the contemporary health care delivery systems with the tools and means by which to reach and support these citizens. AgeTech cannot, on its own, be a work around correcting the dysfunctional characteristics of the delivery system, which must be reformed to better meet the needs of older citizens. AgeTech can be a tool that health care providers can use together with people experiencing poverty to level the playing field providing access, quality and dignity. Reforms capitalizing on and exploiting the benefits of AgeTech are essential, but this must be done without putting the institutions or older adults experiencing poverty at risk. Research, pilots and trials, of the innovations associated with the emerging AgeTech should be encouraged and should be robustly assessed to ensure that they address unmet needs.

Technologies to Bridge the Gap

There are several AgeTech interventions that are demonstrating or showing promising results in helping orphaned older persons, like Francisco, cross the digital divide and integrate more effectively and access health care services.

Highlight Box. Older Homeless Adults

Older homeless adults like Francisco could benefit from portable Internet and phone access. Financial, functional and cognitive impairments are barriers that the homeless often face in accessing healthcare services. Providing older adults experiencing poverty with access to basic technologies could result in improvements to their health status. Programs such as Shelter Tech (https://www.sheltertech.org/programs/) support underserved communities in San Francisco through WiFi connectivity and access to online resources. The program is limited in that provides access to the Internet, but the person still needs access to a smartphone, tablet or computer. During the COVID-19 pandemic TELUS delivered over 10,000 free mobile devices to help isolated seniors, hospitalized patients and at-risk Canadians to stay connected(https://www.telus.com/en/about/news-and-events/media-releases/telus-delivering-over-10000-free-mobile-devices-to-help-canadians-stay-connected). The opportunity for innovators and researchers to work with industry, government, and health care providers, to find ways to put digital technologies in the hands of older adults experiencing poverty is an important imperative to help cross the digital divide.

The Ethical and Practical Imperatives

If AgeTech, digital technology, is to be an effective leveler of access to health care, it must be applied to ameliorate several of the ethical and practical challenges that currently confront the ability of health care delivery system to serve all its citizens equally. The ethical and practical challenges that reforms along with AgeTech ought to be focusing on to help older adults experiencing poverty are (Fig. 9.2).

AgeTech should actively collaborate with older individuals facing poverty, engaging them in the co-creation process with innovators and service providers. This collaborative

- Ageism
- Access to service
- Equitable distribution of health and wealth
- Building trust
- Providing privacy, security, and confidentiality
- Diminishment of power and influence
- Maintaining autonomy
- Assuring quality of care
- Providing value for money

Fig. 9.2 Key ethical concepts related to healthcare

approach will lead to the development of solutions aimed at reducing the barriers that isolate these individuals from essential healthcare services.

Ageism

The WHO labeled the phenomenon of marginalization of older persons as ageism. The Merriam-Webster (2023) online dictionary defines ageism as "prejudice or discrimination against a particular age-group and, especially, the elderly." The older adults experiencing poverty suffer from a double insult, they are not only discriminated against because of their age but also because of lifestyle choices, physical and/or mental disabilities, frailties, addictions, or other marginalizing characteristics. D'cruz and Banerjee (2020, p. 3), in an advocacy review, note that ageism has received an increased profile during the COVID-19 pandemic:

> This is reflected by political leaders and news media in some countries suggesting that the pandemic is a means of culling people who are no longer competent to survive or of use to society (Aronson, 2020). There has also been a call on older adults to stay at home and not burden an overwhelmed health care system—sacrificing themselves for the good of society (Fernandez & Montgomery, 2020). Hashtags such as "boomer removal" have been trending on social media platforms—implying that it is now acceptable to ask older adults to die (Sparks, 2020).

The United Nation's Declaration of Human Rights affords all citizens a life of dignity and fulfilment and the task of this chapter is to shine a light on this segment of the population to ensure that it is extended the same benefits afforded others in publicly and privately funded and delivered health care systems.

Access to Service

Unlike the ethical imperatives and practical challenges identified in the previous chapters where the considerations of older adults are met (or not) within the walls and services of the institution the needs of older adults living in poverty may not receive similar attention. They have the right to the universal health care available to all citizens, but their circumstances are such that they may not avail themselves of the access.

Many take it for granted that Internet access is available ubiquitously, but that is only if you can pay for it to be 'piped' into your home. COVID-19 has shone a glaring light on the strong dependence of our primary health care system on access to the Internet. The Internet is necessary for access to telehealth consultations, monitoring of health status and receipt of laboratory results and other information. If primary health care is meant to be a universally accessible service to all citizens and having access to the Internet is a part of that it follows that people who are experiencing poverty ought to have unfettered access to the Internet. They may, however, not have the knowledge, skills, financial or physical means to access the Internet. There is a dearth of understanding of how at risk older persons are impacted by a lack of access to digital technologies. In an umbrella review, Kunonga (2020) found that:

> The impact of digital technologies on equitable access to services for older people is unclear. Research is urgently needed in order to understand the positive and negative consequences of digital technologies on health care access and to identify the groups most vulnerable to exclusion.

Digital rights have emerged as an important human right that support citizens to have access to computers, other electronic devices and networks (Reglitz, 2020). Health information literacy and access to important health advice may be strong determinants of health. Several countries such as Finland, France, Greece and Spain have extended access to the Internet as a human right.

We propose that Francisco and all citizens have the right of access to the Internet, however, older adults living in poverty also need to have the education, knowledge, skills, support and training to effectively access and navigate the Internet. The primary health care system in the community must ensure that every citizen over the age of 50 will be registered and have access to a primary health care provider, or social worker whose responsibility will be to ensure that they receive the support and health care that they need.

Indigenous peoples generally and older Indigenous persons, are experiencing an increased burden of chronic disease relative to non-Indigenous peoples in many countries. Geographic isolation and segmentation coupled with a history of colonial imperialism linked to intergenerational trauma are contributing factors. Governments and health care delivery systems have put a strong emphasis on working with Indigenous peoples in remote and rural settings to identify ways to provide health care interventions more effectively. Video conferencing, remote biometric monitoring, wearable sensors and smart device apps are now being used to promote health, prevent disease and treat both acute and chronic disease through virtual care. There is, however, a concern with the cultural sensitivity of research and innovation taking place in these communities because of the history of harms from imperialistic intrusions in the past. In response to this the First Nations Health Authority (FNHA) (2023), in British Columbia, have produced a framework on anti-racism, cultural safety and humility.

We propose that the primary health care provider community in collaboration with First Nations communities, researchers and innovators explore ways in which AgeTech can be mobilized to respond to the health care needs of older First Nations peoples' sensitively and appropriately.

Older homeless individuals without access to the Internet are significantly disadvantaged in their ability to access, engage with, and receive health care services. Thurman et al. (2021) in a study of mobile health interventions for homeless persons demonstrated how access to mobile tools improved their ability to self-manage their health successfully. The predominant theme in the study arising from the homeless is that they felt "empowered by technology … how having a smartphone with a plan for unlimited text, calling, data and transportation allowed participants to navigate homelessness and facilitated self-management" (Thurman et al., 2021). Homeless persons that are alienated from the health care delivery system are further disadvantaged by the COVID-19 pandemic.

Making smartphones with cellular voice and data available to the homeless would be a significant step in improving access to health care delivery. Governments and health care delivery organizations must identify how they can develop programs and services that would make smartphones available to those living in poverty and homeless. It has been demonstrated that (Raven et al., 2018, p. 1):

> Older homeless adults could benefit from portable Internet and phone access. However, participants had a lower prevalence of smartphone and internet access than adults aged over 65 years in the general public or low-income adults. Participants faced barriers to mobile phone and Internet use, including financial barriers and functional and cognitive impairments. Expanding access to these basic technologies could result in improved outcomes.

It is important for communities to gain an understanding of the multiple root causes of homelessness. Once understood actions that respond to those root causes should be implemented.

While the root causes are being addressed, we propose that primary health care providers, researchers, innovators and policy makers meet with representatives of the homeless and people experiencing poverty like Francisco to identify their needs and brainstorm ways in which they can fund, distribute, monitor and support (educationally and technically) the homeless in their effective use of the Internet for access to services, information, companionship and decision support. Primary health care providers, policy makers and decision-makers will need to work with the homeless to identify ways in which the smartphones can be funded and provided with a service in which digital literacy and technical supports are provided.

The opportunity for AgeTech innovation in this niche of the technological continuum could be empowering of older adults experiencing homelessness.

Equitable Distribution of Health

An ethical conundrum arising within our health care delivery system is that it is very institutionally oriented around the medical model of care rather than social care. It has been documented that when the public think of public health they think of hospitals and physicians' offices. They do not think of the tools of primary health care and public health such as health promotion, disease prevention, risk reduction, public education or the containment or suppression of a pandemic. The focus of health care delivery is on the downstream consequences of health which manifest as acute and chronic diseases. The pre-eminence and cost intensity of the medical model in our health care delivery system has reduced the role of primary health care to a stature where it does not receive an adequate share of funding and resources necessary to protect and sustain a healthy public, particularly people experiencing poverty. Data on the increasing incidence and prevalence of chronic diseases, obesity, addictions and other lifestyle diseases point toward governments making a strong commitment to using a population health approach to policymaking and funding programs which address inequities in society in the access to and distribution of health care interventions. Digital technologies need to be mobilized to measure, report and monitor the distribution of health and wellness within a society with the view of raising the public's attention to where the gaps and unmet needs are. Government programs would be developed to target the areas of concern with the monitoring and reporting on the outcomes of the interventions.

The COVID-19 pandemic has shone a glaring light on the negative consequences and disease burden that has been experienced by those experiencing poverty in society. A promising development addressing equity is the virtual community for disability advocacy using a searchable artificial intelligence supported platform. This approach addresses the paucity of disability data and information hindering the continuous monitoring of disability equity. In this approach the at risk virtual community uploads and shares validated data and information in support of disability rights advocacy by increasing the communities understanding of the issues (El Morr et al., 2021).

We recommend that AgeTech coupled with health reform be mobilized with research and innovation initiatives to ameliorate the issues of inequitable distribution of health to older adults experiencing poverty.

Privacy and Confidentiality

The conflicting tensions in society between the preservation of privacy and confidentiality of health care data and the expectation of citizens that their health data can be accessible by them and their health care providers at the right place and the right time has been a challenge ever since the emergence of the personal health record and patient portals. Several countries with effective primary health care systems such as Denmark, Finland and the Netherlands have been exemplary in the adoption and provision of electronic health records to the population but many, like, Canada have not. They have been laggards in adoption, diffusion, and utilization. We expand on this challenge in Chap. 8.

AgeTech must come to the aid by forging and sustaining a way forward on the ethical conundrum of providing important information and simultaneously preserving privacy and confidentiality. For example, the use of locator services with global positioning system (GPS) can provide a family of a patient with early-stage dementia with an opportunity for them to live at home independently but it raises the question of invasion of privacy to have their movement and location monitored. The tension between the values of an individual's autonomy and preservation of their security can come into conflict.

The way forward on these issues is to recognize that the health data of a citizen is owned by that individual, and they and their families decide how it is to be shared and used. Today in many jurisdictions' governments, institutions, clinics and private providers are required by law to act as custodians of a citizen's personal health record. These laws and regulations were set in place before there was the technical capability to provide citizens with a way to easily access and control the use of this information. This is no longer true and there is a need for modernized legislation and regulations coupled with AgeTech to enable and empower citizens to own, access, protect and control who can have access to their health information.

A Pan-Canadian health data charter was introduced in 2023 committing the governments to adopting common standards and policies so that citizens across Canada can easily access their health information and benefit from having it securely shared across health care settings while protecting their privacy. The charter requires "Person-centric health information design to ensure that health data follows the individual across points of care to support individual, clinical, and analytical access and use while respecting individual privacy…". (Government of Canada, 2024). The national and provincial governments are currently establishing mechanisms to advance their efforts on this important initiative.

We propose that primary health care providers and AgeTech be mobilized to provide older adults living in poverty like Francisco with the equipment, knowledge, skills and support to give them ownership of their longitudinal health and personal health record with the view to providing them with control and access to the full range of types and levels of health care services.

Governments collect and publish data on the number of substance abuse and opioid related deaths that take place within our jurisdictions, but we do not know how many of those are older adults living in poverty. We have rough data estimates of the numbers of homeless people in our communities, but we have little accurate data or understanding of their needs for health care and social assistance. We do not know very much about the experiences, magnitude or trends associated with the problems this cohort of citizens face. We lack data and understanding that tells us the number of people experiencing poverty in our communities, what risks they are experiencing, where they are located and how we can reach out to them. There is a recognition that the boundary between individual privacy and autonomy and state interference must be respected but we have not carefully identified and navigated effective ways to mitigate that territory to ensure that older adults experiencing poverty lead a dignified and fulfilling life. It is difficult for a community or society to recognize and respond to this population unless we have a better understanding of their plight through data and information.

In response to this dearth of understanding Sacco et al. (2023) undertook a qualitative study to examine the experiences of health care providers who work with and support older adults experiencing poverty. Sacco et al. (2023) identified several system level issues, "limited funding, insufficient resources and supports tailored to an older homeless population, coordinating care with different sectors, the cyclical nature of homelessness, and stigmatization" (Sacco et al., 2023). Sacco and colleague's findings reinforce the lesson that health care providers' experiences in providing support to older adults experiencing poverty can be an important ingredient in the co-creation of strategies to help address the unmet needs of this population. Thinking about Sacco and colleague's findings in the context of AgeTech we encourage innovators to identify means by which digital technology can be co-created and used to build and sustain a strong relationship between the provider and the aging adult experiencing poverty. These include such undertakings as building trust, encouraging client choice and adopting central support roles because their clients have lost contact with their families, neighbors or other people that would have been important in their lives. Health care providers, within the bounds of their professional scope of practice, may take on the role of a family member figure to support a client. This may involve roles such as making appointments, participating in consultations, booking transportation, finding shelter, completing income assistance documentation and managing medications (Sacco et al., 2023). AgeTech applications that support health care providers in fulfilling these roles would be of great value to the clients and the providers.

A second finding recognizes the importance of acknowledging the 'whole person' and moving from responding to an immediate crisis to providing a bridge to health care required on a sustained basis to treat a chronic health, mental health, or addictions issue. AgeTech applications that provide an opportunity to facilitate an ongoing and sustained collaboration among the clients, their team of care providers and their community partners would expedite the continuity of care and support. Sacco found that the sharing of information between providers in multiple care and community settings is often disjointed and fragmented leading to a lag or termination of supports. Sacco et al. (2023) identify further exacerbations from "limited funding, coordinating care with different sectors, inadequate resources and supports tailored to an older homeless population, the cyclical nature of homelessness, and stigmatization" (Sacco et al., 2023). AgeTech innovations that respect privacy and confidentiality but harness the opportunity to facilitate a smooth flow of information among organizations would significantly improve the effectiveness of care providers' service to their clients.

Older adults experiencing poverty are confronted with challenges navigating the health care delivery system, "through three main community transitions (1) correctional facilities into the community; (2) hospital into the community; and (3) community hospital into long term (LTC)" (Sacco et al., 2023). Remedial interventions of system reform coupled with AgeTech could involve improved discharge planning for transitions between different types and levels of care and support. Voluminous paperwork is identified as another barrier that could be accommodated through the application AgeTech solutions that have been co-created with input from clients and field tested through usability studies to ensure that they are a time and resource saving tool.

Building Trust

Older adults experiencing poverty are rarely involved in the co-creation and advancement of AgeTech for the purpose of meeting their specific health care needs. Researchers in a study seeking to understand the practical and ethical issues of conducting research into the use of AgeTech in people's homes identified four key themes that determined how comfortable they felt with the research:

- motivation for participating
- transparency, understanding and consent
- privacy, anonymity and data use
- trust in research

These findings are important in identifying practical ways in which AgeTech research can be facilitated.

The factors that were relevant to developing and maintaining this trust were the trustworthiness of the research team, the provision of necessary information, participants' control over their participation, and positive prior experiences of research involvement (Kennedy et al., 2021).

Trust plays an important role in maintaining a positive relationship among researchers, innovators and the public.

We propose that older adults experiencing poverty like Francisco be enabled and empowered to work with researchers, providers, policymakers, industry and government to develop a policy framework and statement on cultural safety and humility for researchers and innovators working with them.

Diminishment of Power and Influence

Between the beginning of the pandemic in March 2020 and March 2022, well over 20,000 older Canadians' deaths were attributed to COVID-19 in residential LTC facilities (Loreto, 2023). Canada performed relatively poorly in protecting its older clients in LTC facilities. A CIHI report cites Canada as a poor performer in an international peer group comparison (CIHI, 2023). Another report by the Office of the Seniors Advocate in British Columbia (2021) states that LTC residents would have chosen to have lived at home rather than an institution during the pandemic. Older adults experiencing poverty who did not live in a LTC facility during the pandemic, still had a different set of barriers and challenges facing them as their age and co-morbidities contributed to their greater risk of illness and death. This issue underlines the ethical issue of the diminishment of the power in the voice of older persons. Ageism is a serious challenge but particularly so for older adults experiencing poverty. We advise that the older adults living in poverty be encouraged and motivated to help co-create ways and means to innovate AgeTech to increase their political power and influence.

The older victims of domestic violence or elder abuse are often restricted in their ability to break out of relationships that are emotionally and or physically harmful to them. AgeTech coupled with other needed reforms may provide avenues and modes of support that could not otherwise be accessed from our health care and delivery systems. As with any technology there are both benefits and risks.

We propose that the primary health care providers and the AgeTech community collaborate to explore and mobilize the use of effective digital technologies to provide older adults experiencing poverty like Francisco with tools to empower their voices and influence in society.

Autonomy, Safety and Security

The social media milieu presents a delicate balance between the preservation of privacy and security of the individual in their relationship with society and the health care delivery systems specifically. The ascendency of digital technologies has had many positive results for social and cultural richness worldwide, however they are also implicated in facilitating spread of misinformation, violence, crime and abuse. At a global level cybercrime is increasing and at a personal level hate speech, scams and online harassment are increasing in frequency. Older adults are often targeted, and older adults living in poverty are at a greater risk. In a recent Pew Research Center report (2021), *The future of digital spaces and their role in democracy,* experts expect that the negative effects from digital tools and forums that exploit people's frailties, stoke their rage, and divide them will continue in the short term but that democracies will need to respond with initiatives to facilitate interaction with enhanced civility and assured personal security (Anderson & Rainie, 2021). AgeTech can play a significant role in identifying ways to involve older adults living in poverty in identifying ways that their autonomy, safety and security can be preserved and at the same time providing the benefits of addressing their unmet needs.

We propose that older adults living in poverty, like Francisco, in concert with governments, seniors' advocacy organizations, law enforcement, health care providers, researchers and innovators identify programs and funding that will incentivize the development, assessment and diffusion of digital technologies that help address the issues of elder abuse and cybercrime directed toward older adults experiencing poverty.

Quality of Care and Value for Money

A further ethical issue that arises for older adults experiencing poverty is the quality of the care that they receive from the primary health care system. The literature is replete with case examples and observational studies documenting the frustration that older adult patients experience in their interface with the health care delivery system. The challenges for the older adults are significant but the aging adult experiencing poverty is even further disadvantaged. Whether the issue is being put on hold to make an appointment, dropping a call, interacting with an automated message service, waiting an inordinate amount of time to speak to a health care provider, not having ubiquitous access to an EMR when it is needed, or relying on an antiquated form of communication (fax), these are the daily challenges they face. In addition to these issues of convenience are the primary health care outcomes from interventions that have been demonstrated to be superior if they are in a location of the person's choosing for example, the home, rather than in the institutional setting.

We propose that primary care providers and the AgeTech community be incentivized and mobilized to provide universal access for older adults experiencing poverty, like Francisco, with a smart device that allows them link to their primary care provider or social worker so that the two of them can monitor their health care status and outcomes from health care interventions on a real-time basis. Quarterly summaries and annual reports can be produced and shared between the citizens and the primary care provider to ensure that health status is monitored and adjusted to meet the needs of the individual.

A further area of concern is that of the 'promised' outcomes or value to the older adults experiencing poverty in the form of the digital technology that has been developed. Digital devices that sound good in marketing brochures and media advertising but are weak on delivery of tangible benefits for the older adults experiencing poverty are a significant concern. This is a problem not only for lack of utility of the digital technology produced but also the opportunity cost associated with the foregone funds that could have been used toward better ends. Industry associations have been sceptical about the added value of rigorous standards to demonstrate effectiveness because they perceive it as an inappropriate barrier to market access. However, MedTech Europe have agreed and demonstrated that some regulations are necessary (MedTech Europe, 2023).

Isabel Van De Keere, program director at the Digital Medicine Society (DiME), said: "EU regulators just took a major step forward towards the harmonisation of health technology assessments across EU member states with positive implications for health care technology innovators and the public in need of safe and effective health and care services. As we saw throughout the pandemic, collaboration between health care regulators, policy makers and healthcare technology innovators across regions is the only way forward towards an accessible, effective and equitable healthcare system" (Lovell, 2021).

We propose that primary health care providers along with their roster of older adults experiencing poverty like Francisco participate in the pre-market development, post market assessment, surveillance and reporting of the appropriateness and effectiveness of AgeTech digital technologies that are jointly developed by the researchers, innovators and industry. This incentivizes the continual improvement of the digital technologies in support of older adults experiencing poverty.

AgeTech promises to make significant contributions to improving the health care delivery systems and in turn meeting the health and social care needs thereby improving the quality-of-life of older adults experiencing poverty. AgeTech however cannot do this alone—significant policy and practice reforms are necessary to facilitate the creation and adoption of AgeTech that reflect today's societal values and technical capability. By providing older adults living in poverty control of responding to their conditions to raise them from their states of vulnerability we are giving them the opportunity to identify how they can mobilize AgeTech to help themselves.

We must prepare the health care delivery systems for the future with a framework and approach that removes the societal, legal, policy, administrative and practice barriers that magnify their vulnerability and focus instead on those innovations with AgeTech that will lift them from their vulnerability.

Rubric for Advancing AgeTech to Address Unmet Needs

Sacco et al. (2023) identified six primary roles that health care providers had to fulfill in their support of homeless older adults: educator, health support, social support, advocate, collaborator, system navigator. If we take the roles and cross tabulate them with the ethical imperatives that require a health reform or AgeTech solution in Table 8.1 is the result. The opportunity for innovators and researchers is to fill in the cells with both a system reform coupled with an AgeTech solution. The roles and imperatives within the cells must not be interpreted as discrete elements but rather interconnected to a larger whole where a continuity of structures and functions must work seamlessly and in support of one another (Table 9.1).

Researchers, innovators, industry, health care providers and policymakers are encouraged to use a "whole person" approach and collaborate with older adults experiencing poverty to arrive at AgeTech solutions that will respond to their unmet needs. This approach can also apply to the broader citizen population and does not need to be focused on the people experiencing poverty in society.

Table 9.1 AgeTech opportunities matched to ethical imperatives by role

Role/ethical imperatives	Educator	Health support	Social support	Advocate	Collaborator	Policymaker
Access						
Equity						
Privacy						
Trust						
Power/ Influence						
Autonomy/ Safety						
Quality of care						

Note Adapted from Sacco (2020). A missing piece of the puzzle: Health care providers' experiences supporting older homeless adults in outreach settings. Master's thesis, University of Waterloo, UWSpace. https://Uwspace.uwaterloo.ca/handle/10012/16573

Seminal Demonstration Project

The COVID-19 pandemic underlined the vulnerability of older persons' access to their families and friends all due to their lack of access to the Internet. We propose a seminal longitudinal demonstration project over a decade to demonstrate innovative bridging the digital divide of older adults experiencing poverty. The project would provide older adults with digital access to their community and the health care delivery system. It is proposed that the project represent older adults experiencing poverty persons from urban, rural and remote locations, a representative sample of the ethnic, cultural and racial mosaic of the jurisdiction's population, and identify a heterogenous range of economic, clinical, psychological and social vulnerabilities of the older persons. The project would provide at risk older persons with an Internet ready portable device with a connection to the Internet. The seminal demonstration project structure, program development and evaluation would be co-created with representative at risk older persons. The older persons with portable devices would be supported with education, hands on training sessions and a 7 days a week 24 h a day help line. Implementation would include instruction to individuals on a real-time basis as to how to use the device to seek information, make appointments, carry out teleconsultations, retrieve test results and connect with family and friends. It is proposed that the seminal demonstration project have a continual and comprehensive audit and review process that monitors and identifies the challenges and issues associated with bringing the older persons online. Remedial action would transpire immediately on the identification of issues and challenges. The learning system approach would be utilized to structure the project. The objective would be to identify ways and means to cross and sustain a bridge across the digital divide experiences of at risk older persons in the community.

Concluding Remarks

The health care systems and the primary care systems within them in many countries are digitally challenged and are still operating at rudimentary stages. Many of the legislative, regulatory, policy and health care practice and delivery approaches of the past have surpassed their 'best before date.' As demonstrated in this chapter, the structures of the past, which have extended health care to so many, have become the Achilles' heel of today limiting the opportunity for progress by reaching out to the older adults experiencing poverty and provide them the same access and support those other citizens enjoy. Policymakers leading health care delivery reforms must partner with older adults living in poverty to identify and develop effective means by which the innovation with AgeTech can be incentivized to address the digital divide. This will require due diligence to ensure that the innovations are truly effective in delivering on their promise as well as providing citizens with value for the money expended. This will require a strong partnership of older adults experiencing poverty, government, researchers, innovators, industry and providers.

The reader is encouraged and provided with the following observations and recommendations as to how to accelerate and sustain the effective introduction of AgeTech to older adults experiencing poverty:

- Make mandatory and barrier free universal access to smart devices to all at risk and older adults experiencing poverty providing them with unfettered access to the Internet and in turn the health and social care delivery systems.
- Provide education, training and support to the older adults experiencing poverty population to effectively use smart devices to access and maintain support from the care providers, monitor adverse incidents, provide status reports and request elective, urgent or emergent health care or social services as needed.
- Develop, assess and refine platforms, programs, supports and apps co-created with older adults experiencing poverty for the smart devices to facilitate access to shelter, food, clothing, social community connections, their health and social care services, data and information, etc.
- Integrate the older adults experiencing poverty in identifying the unmet needs of their health and social care circumstances that AgeTech can be mobilized to help address and resolve. Co-create solutions.
- Integrate and coordinate the legislative, regulatory, policy, funding and program delivery of social and health care delivery supports to the older adults experiencing poverty.
- Provide the older adults experiencing poverty and homelessness with ownership, access and control of their health information.
- Implement a secure and confidential digital health care record to international standards of best practice that provides all health care and social care providers with the information synchronously.
- Develop an anonymized secure and confidential national (CIHI) nomenclature and longitudinal database on the prevalence, incidence, status, health and social care utilization, adverse incidents, trends and outcomes associated with older adults experiencing poverty.
- Synthesize the anonymized data on older adults experiencing poverty so that it is available to researchers and innovators to explore and identify ways to support them.
- Educate and train researchers, innovators and providers in cultural sensitivity and humility before inviting them into the community of older adults experiencing poverty and ask them for advice and direction for what needs must be addressed.

References

Alkureishi, M. A., Choo, Z., Rahman, A., Ho, K., Benning-Shorb, J., Lenti, G., Velazquez Sanchez, I., Zhu, M., Shah, S. D., & Lee, W. W. (2021). Digitally disconnected: Qualitative study of patient perspectives on the digital divide and potential solutions. *JIMR Human Factors, 8*(4). https://doi.org/10.2196/33364

Anderson, J., & Rainie, L. (2021). *The future of digital spaces and their role in society.* Pew Research Center.

Aronson, L. (2020). Ageism is making the coronavirus pandemic worse. *The Atlantic*, 28. https://www.theatlantic.com/culture/archive/2020/03/americas-ageism-crisis-is-helping-the-coronavirus/608905/.

Baah, F. O., Teitelman, A. M., & Riegel, B. (2019). Marginalization: Conceptualizing patient vulnerabilities in the framework of social determinants of health—An integrative review. *Nursing Inquiry, 26*(1), e12268. https://doi.org/10.1111/nin.12268

Canadian Institute for Health Information (CIHI). (2023). *Long-term care and COVID-19: International comparisons.* CIHI. https://www.cihi.ca/en/long-term-care-and-covid-19-international-comparisons

D'cruz, M., & Banerjee, D. (2020). 'An invisible human rights crisis': The marginalization of older adults during the COVID-19 pandemic–An advocacy review. *Psychiatry research*, 292, 113369. https://doi.org/10.1016/j.psychres.2020.113369

El Morr, C., Maret, P., Mehlenbach, F., Dharmalingam, D., Tadesse, R., Creighton, A., Kundi, B., Buettgen, A., Mgwigwi, T., Dinca-Panaitescu, S., Dua, E., & Gorman, R. (2021). A virtual community for disability advocacy: Development of a searchable artificial intelligence-supported-platform. *JMIR Formative Research, 5*(11), e33335. https://doi.org/10.2196/33335

Fernandez, M, Montgomery, D. (2020). Texas tries to balance local control with the threat of a pandemic. N. Y. https://www.nytimes.com/2020/03/24/us/coronavirus-texas-patrick-abbott.html.

First Nations Health Authority British Columbia, Anti-Racism, Cultural Safety & Humility Framework. (2021). https://www.fnha.ca/Documents/FNHA-FNHC-FNHDA-Anti-Racism-Cultural-Safety-and-Humility-Framework.pdf

Government of Canada. (2024). *Pan-Canadian health data charter.* https://www.canada.ca/en/health-canada/corporate/transparency/health-agreements/shared-health-priorities/working-together-bilateral-agreements/pan-canadian-data-charter.html

Hall, M. M., Stevens, P. E., & Meleis, A. I. (1994). Marginalization: A guiding concept for valuing diversity in nursing knowledge development. *Advances in Nursing Science, 16*(4), 23–41. https://doi.org/10.1097/00012272-199406000-00005

Kennedy, M-R., Huxtable, R., Birchley, G., Ives, J., & Craddock, I. (2021). A question of trust and a leap of faith—Study participants perspectives on consent, privacy, and trust in smart home research: Qualitative study. *JIMR Mhealth and Uhealth, 9*(11). https://doi.org/10.2196/25227

Kunonga, T. P., Spiers, G. F., Beyer, F. R., Hanratty, B., Boulton, E., Hall, A., Bower, P., Todd, C., & Craig, D. (2021). Effects of digital technologies on older people's access to health and social care: Umbrella review. *Journal of Medical Internet Research, 23*(11), e25887. https://doi.org/10.2196/25887

Loreto, N. (2023). *Deaths in Canadian residential care facilities.* [Spreadsheet]. https://docs.google.com/spreadsheets/u/0/d/1M_RzojK0vwF9nAozI7aoyLpPU8EA1JEqO6rq0g1iebU/htmlview

Lovell, T. (2021). *European council gives the go-ahead to new healthtech assessment rules.* https://www.mobihealthnews.com/news/emea/european-council-gives-go-ahead-new-healthtech-assessment-rules?utm_source=eHealthrecord&utm_campaign=7a729a1ca4-

MedTech Europe. (2023). *MedTech Europe welcomes the adoption of amended transitional provisions of the Medical Devices Regulations and calls for continued work to address outstanding implementation challenges*. MedTech Europe. https://www.medtecheurope.org/new-medical-technology-regulations/

McDonald, L., Dergai, J., & Cleghorn, L. (2007). Living on the margins: Older homeless adults in Toronto. *Journal of Gerontological Social Work, 49*(1–2), 19–46. https://doi.org/10.1300/J083v49n01_02

Office of the seniors advocate British Columbia. (2021). REVIEW OF COVID-19 OUTBREAKS IN CARE HOMES IN BRITISH COLUMBIA, https://www.seniorsadvocatebc.ca/app/uploads/sites/4/2021/10/Outbreak-Review-Report.pdf

Pew Research Center, November (22, 2021). The future of digital spaces and their role in democracy https://www.pewresearch.org/wp-content/uploads/sites/20/2021/11/PI_2021.11.22_2035-Digital-Public-Sphere_FINAL.pdf

Raven, M. C., Kaplan, L. M., Rosenbert, M., Tieu, L., Guzman, D., & Kushel, M. (2018). Mobile phone, computer, and Internet use among older homeless adults: Results from the HOPE HOME cohort study. *JMIR Mhealth and Uhealth, 6*(12), e10049. https://doi.org/10.2196/10049

Reglitz, M. (2020). The human right to free Internet access. *Journal of Applied Philosophy, 37*(2), 314–31. https://doi.org/10.1111/japp.12395

Sacco, V., Stolee, P., MacEachen, E., & Boscart, V. (2023). Canadian health care providers perspectives on working with older homeless adults in outreach settings. *Canadian Journal on Aging*. Accepted for publication.

Sparks, H, (2020). Morbid'boomer remover' coronavirus meme only makes millennials seem more awful [Internet]. N. Y. Available from: https://www.nypost.com/2020/03/19/morbid-boomer-remover-coronavirus-meme-only-makes-millennials-seem-more-awful/.

Thurman, W., Semwal, M., Moczgemba, L. R., & Hilbelink, M. (2021). Smartphone technology to empower people experiencing homelessness: Secondary analsyis. *Journal of Medical Internet Research, 23*(9), e27787. https://doi.org/10.2196/27787

A Vision for Health Care and AgeTech in the Twenty-First Century

10

Introduction

The global community of health care delivery systems exemplify the full range of program delivery for older adults from barely adequate to close to ideal. No one delivery system is ideal. They all have elements of excellence in certain sectors and weaknesses in others. In this chapter we reinforce the importance of co-created and co-developed reforms in the legislative, regulatory and policy landscapes that shape the structure and functioning of health care systems and the degree to which AgeTech can be a positive factor in those reforms. The reader is introduced to the new and emerging values that citizens, particularly older persons expect from their health care delivery systems. The chapter highlights conceptual descriptions of the health care systems and an idealized system of health care delivery that optimizes an older person's autonomy, dignity and zest for life (Fig. 10.1).

The Role and Importance of Reforms in the Legislative and Policy Landscapes

We have chosen a broad definition of AgeTech in this book to provide the reader with the broad range of legislative, regulatory, funding, policy and practice challenges and opportunities that affect the ability of AgeTech to be fully optimized for the benefit of older adults. Many countries are laggards and Canada is a particularly poor performer in the co-evolution and integration of AgeTech into the health care delivery system (Webster, 2015). This means that the backbone and required infrastructure such as access to the Internet and the ubiquity of the electronic health record are not accessible to citizens in different settings. Governance issues of who owns the patient data, for example, the

D. Juzwishin et al., *AgeTech Innovations in Healthcare for Older Adults*, Synthesis Lectures on Technology and Health, https://doi.org/10.1007/978-3-031-86885-6_10

Fig. 10.1 Technology bridging health and home. *Credit* Age without limits—age-positive image library (https://www.agewithoutlimits.org/image-library)

physician's office, the hospital, or the patient, further exacerbate the tension of making progress. Harm comes to older persons as a result of health data not being interoperable and shared among health data bases.

Many health care systems exhibit a resistance to innovation, not because innovation is not valued, but because the legislative, policy and practice threshold for risk taking is very low. If public funds are expended on an innovation that does not prove value, public embarrassment may arise for policymakers or providers eroding the confidence in decision-makers or policymakers. This inability to develop business practices that demonstrate the value of an innovation transparently and deliberately inhibits the adoption of AgeTech. Mature policy and procurement practices need to be developed that will permit the legitimization of some degree of failure being part of the innovation process.

David Naylor was commissioned by the Government of Canada to report on a multi-year agenda that could help transform Canadian research and innovation capacity (Advisory Panel on Healthcare Innovation, 2015). Naylor, in the report *Advisory panel on healthcare innovation, 2015*, identified the following as barriers to innovation:

- lack of meaningful patient engagement
- outmoded human resource models
- system fragmentation
- inadequate health data and information management capacity

- lack of effective deployment of digital technology
- barriers for entrepreneurs
- a risk-averse culture
- inadequate focus on understanding and optimizing innovation.

The aversion to risk in the Canadian and other health care systems is baked into the laws, regulations, policies, values, practices and funding models that maintain the status quo. This characteristic has been documented and confirmed in a recent scoping review by MacNeil and colleagues (2019):

> The influence of the Canadian policy context found in this study aligned with other international findings that point to the influence of macro-level factors such as political structures and macroeconomic and fiscal policies on health innovation diffusion. Although micro level factors did not emerge strongly in this scoping review, others have suggested a focus on the culture at the front lines of health care, which may be more amenable to intervention than macro system factors.

In response to this weakness in health care systems the authors depicted the stages along the innovation pathway and the crosscutting influence of a health care system's policy context, resources, and partnerships and communication on technology development, assessment and implementation.

MacNeil et al. (2019) made the following observations:

> In addition to these stages, we are aware of emerging areas within the health technology assessment literature, which emphasize the importance of evaluating health technologies over their life cycle. Ongoing evaluations and delayering innovations plays an important role in creating budget flexibility to support adoption of new technologies, and a sustainable system over time. Rather than ending at implementation, the innovation pathway requires sustainability. Another emerging theme is the recognition and current effort focused on engaging users in co-creating relevant technologies. The meaningful engagement of patients and caregivers in the development and adoption of useful innovations has been echoed internationally and regional innovation ecosystems have been proposed as a mechanism through which to engage these users to ensure technologies are aligned with health system needs. Although outside the scope of our search, we have incorporated these findings into a revised graphic depiction of Canadian health technology innovations and support an ongoing emphasis of engagement of users throughout the innovation process.

The path dependency exhibited in many health care delivery systems arises from legislation, policy, funding models and program delivery creating a natural aversion in the systems to be innovative.

Lessons from the Canadian example may be applied to the broader global community that the client or patient should be situated and interacting during the entire iterative cycles of development, assessment, implementation and sustainability of the innovation. The cycles are sustained throughout the policy context with support of resources and partnerships focused on crosscutting themes. Specific to the Canadian context is the framework in the diagram, however, lessons can be applied to the broader global community that the client or patient should be situated and interacting during the entire iterative cycles of development, assessment, implementation and sustainability of the innovation. The cycles are sustained throughout the policy context with support of resources and partnerships focused on crosscutting themes.

An Idealized Health Care System

Transitions between and among the settings and the types of care although identified as discrete entities are expected to be facilitated seamlessly through transportation, communication, staff support, information exchange along with other necessary logistical factors. This underlines the important role of primary care (Chap. 7) as the coordinator and facilitator of care and support transitions, particularly when it comes to addressing the challenges of those older adults who fall through the cracks (Chap. 9). We do not identify other sectors of the health and social care systems such as acute care hospitals, group homes, income support and public health, however, they are important contributors and partners with linkages and exchange that must be accommodated. The scope of this book does not permit identifying the linkages and exchange with those programs, but our concern is that those be expedited as well.

Models can never reflect reality and the movement of clients between and among the types and levels of care. For example, continuity to acute care although important is not captured. In addition rehabilitation can take place in all of the settings.

The Life Trajectory Approach

The approach taken in this book follows the individual in their life trajectory and transitions from one health state to another and, therefore, from one set of health and social needs to another in the later stages of life. These trajectories have been represented as *personas within scenarios*. This approach ensures that we do not view an individual as existing within a static state without the opportunity of having them stretch toward the potential of maximal independence. In this book we emphasize the interaction and relationships among recipients of support and care, providers, innovators and technological products. We envision the development and acceleration in adoption of AgeTech, see first diagram, as not being restricted to a linear process of disjointed and incremental staged development but rather an iterative looped process with multiple branches of innovation and experimentation in search of continual refinement to meet immediate needs as well as anticipating those that will arise in the future. AgeTech solutions may need to be customized for everyone to meet their specific needs.

We identify and analyze the appropriateness, effectiveness, and efficiency of technologies from the perspective of older adults, their families, communities, health care providers and institutional settings. We also take the position that the current dysfunctional structural characteristics and processes of the health and social care systems must be reformed and that technology is not used to provide temporary solutions because that would only perpetuate the dysfunctions. AgeTech is meant to create a fulfilling life for older adults and not to act as workarounds for dysfunctional organization issues. This is not to say that it cannot be a two-way street. What benefits the client can be tailored so that it helps organizations deliver more effective and efficient care. We advocate for reform and for addressing those inadequacies. This book is culturally and ethnically sensitive as it recognizes that one solution may not fit all nuances of individual values and beliefs. When technology effectively serves the diverse needs of all participants in the care space it can be said that the adoption of the technology is successful.

AgeTech is the most successful when it is critically assessed and creates the necessary ingredients coupled with synergies in supportive environments where mutual aid and self-care optimize individual autonomy and achievement of a fully potentiated quality-of-life. The reader is encouraged to think about these interactions and interconnections as they co-create AgeTech innovations with older persons and providers.

Supportive environments are central to health care services delivery, creating and nurturing policies, services, practices, programs and surroundings that enable healthy aging in the settings where older persons live, work, learn, love, create and worship. They reach out beyond the health sector and link to transportation, housing, and fiscal policy. These are dealt with in other books in this series.

Mutual Aid, represents relationships and actions of people, such as families, neighbors and friends to support each other emotionally, physically and mentally by sharing ideas, information, resources and experiences.

Self-care, refers to the abilities, choices and actions individuals take in the interest of their own health such as choosing to be active, join a community organization and to safety proof their home.

Readers are encouraged to think about and act on the conceptual frameworks represented in this chapter to catalyze a synergistic relationship between necessary health care reforms and successful innovation with AgeTech. In the next section of this chapter, we will identify and review the values that are core to respecting an older person's enjoyment of a life with dignity and zest.

AgeTech, Aging and Values

An unbridled, uncritical and undisciplined approach to the successful development and application of technologies, and in this case AgeTech, promising to improve the quality-of-life is rarely successful unless it actively integrates values of older persons. More often the negative or unforeseen consequences have a detrimental effect for both the persons exposed to the technology and for that society's support for research and innovation. AgeTech is particularly susceptible to this because older persons may exhibit characteristics of reduced physical dexterity, sensory perception and cognitive function. This places an even stronger emphasis on the emergence of AgeTech to be sensitive and respectful of these natural processes accompanying aging. Financial resources in society are constrained, therefore the opportunity cost associated with misspent funds on innovations that do not materialize in quality-of-life improvements for citizens means that an alternative societal benefit may not be realized. For this reason, we promote a disciplined approach to financially stimulating and incentivizing the creation, development and implementation of AgeTech by relying on the approaches and methods of health technology assessment (Juzwishin et al., 2018).

Contemporary health technology assessment and innovation is anchored in four fundamental ethical principles or values that we advise researchers, innovators and health care providers to keep in mind when assessing how to respond to an unmet need with a digital technology. In this way we can ensure that the digital technology is in the service of humanity rather than for its own sake. We refer to these as the ethical imperatives that reform coupled with AgeTech must be prepared to respond to.

Highlight Box. Digital Health Evaluation Network

Canada's drug and health technology agency, Canadian Agency for Drugs and Technologies in Health (CADTH), has established a digital health evaluation network (*Centre for Digital Health Evaluation* (CDHE)). The network was established to facilitate the production of evidence on the impact of digital health interventions on access, quality, and equity of health care delivery. To address the issues of uncertainty around effectiveness, comprehensive and clinically relevant evaluations are needed. The CDHE will incorporate researchers, decision-makers, policymakers, patients and other stakeholders to facilitate knowledge translation and mentorship to build capacity in this emerging field.

The first value is that AgeTech ought to *respect* and *facilitate* the desire of individuals to *live* and *thrive independently* in their own home or community surrounded by family, friends and neighbors for as long as possible or is feasible. AgeTech needs to respond to unmet needs. It should not create tools that do not add value to a rich and fulfilling life. During the COVID-19 pandemic health technology assessment agencies and researchers shifted their focus toward knowledge translation and implementation issues surrounding the effective use of telemedicine and other technologies because of the restrictive nature of dealing with the pandemic (Cooper-Jones et al., 2022).

The second value is that AgeTech ought to be *co-created* with the older adult, not independently of them. Older adults need to be brought into the conversation about what their needs are and how they are perceived and understood. An older person manifesting frailty will have different needs for a remote control for their television than would an able-bodied individual. The entire journey from idea generation to invention, proof of concept, prototype development to trialing should be conducted with the older person as a meaningful partner as illustrated in the diagrams. We also caution that AgeTech not assume that the institutional or provider view is representative of what the older adult may want or need. Validate your assumption(s) with the older adult. AgeTech is the concierge to the client.

A third value is that AgeTech ought to *facilitate and encourage the smooth transition* through the life cycle with a view toward maximum independence, always bearing in mind the balance among autonomy, quality-of-life and safety of the individual. Hence, we are identifying technological supports and services that maintain a high quality-of-life *and* expedite the smooth transition between and among different types and levels of care, always with the highest level of practical independence as the goal. The imperative of the health and social care system ought to be to ensure that citizens are able to access the right type of support and care at the right location, from the right service provider at the right time. Path dependency and boundary maintenance of organizations working in 'silos' protecting their 'turf' are major barriers to the provision of smooth care. We prescribe and encourage ways that AgeTech can help facilitate continuous care delivery while protecting the security, privacy and confidentiality of citizens. This will involve identifying and exploring ways in which families and community can be mobilized through communication coupled with informatics platforms and apps to wrap themselves around those who need care and support.

A fourth value is for *the education and preparation of students and researchers* in the fields of engineering, computer science, health information science, health care provision and leadership to mobilize a client centered approach to co creation of AgeTech innovations by having students do placements in the settings to understand first-hand the challenges older adults may face. Bridges between and among clients in need, industry, academia, research and practice must be built to maximize the potential of each contributing member. For example, AGE-WELL, a Canada-wide initiative, is preparing trainees and early career researchers to be leading innovators in digital health solutions for older persons. The program equips graduate students, postdoctoral fellows, and early career researchers to facilitate the adoption and implementation of AgeTech for older persons with complex needs and their caregivers.

Table 10.1 Checklist of core values to drive AgeTech

No	Value	Description	Y/N
1	Independence autonomy agency	Respect and facilitate the desire of individuals to live and thrive independently in their community	
2	Co-creation	Co-create with the older adult not independently of them	
3	Facilitate transitions	Facilitate transitions through the stages of life toward maximum independence, balancing autonomy, quality-of-life and safety of the individual. Remove barriers, challenges and disincentives that prevent the smooth transition of citizens and their data between and among the types and levels of care. Do not use AgeTech as workarounds or crutches to tolerate institutional barriers	
4	Increase knowledge, understanding, skills	Educate and prepare educators, students and researchers to mobilize a client centered approach to co creation of AgeTech innovations by having students do placements in the settings to understand first-hand the challenges older adults face	

The following checklist provides, for health care providers, researchers, innovators, educators, policy and decision-makers, a list of core values to ensure that their efforts reflect the core values of AgeTech (Table 10.1).

Concluding Remarks

An opportunity for today's and future AgeTech is to restore and maintain people living independently, however most technologies are geared toward able-bodied individuals. Individuals who are compromised in their cognitive, physical, visual, auditory, tactile, taste, speech or haptic senses cannot effectively use many technological aids on the market. When severe cognitive or mobility issues arise, additional barriers occur, many of which appear to be insurmountable. Many technologies that potentiate independence do not exist or are still in the early stages of development. We identify promising opportunities for researchers, innovators, educators, providers and industry to pursue. We are committed to and promote the values of autonomy, equity and unfettered access to life enhancing AgeTech.

References

Advisory Panel on Healthcare Innovation. (2015). *Unleashing innovation: Excellent healthcare for Canada. Report of the advisory panel on healthcare innovation.* Government of Canada, Health Canada. Health Canada. https://healthycanadians.gc.ca/publications/health-system-systeme-sante/report-healthcare-innovation-rapport-soins/alt/report-healthcare-innovation-rapport-soins-eng.pdf

AGE-WELL. (2023). *About AGE-WELL.* https://agewell-nce.ca/about-age-well

Batalden, M., Batalden, P., Margolis, P., Seid, M., Armstrong, G., Opipari-Arrigan, L., & Hartung, H. (2016). Coproduction of healthcare service. BMJ Quality & Safety, *25* (7), 509–517.

Bryan, S., Mitton, C., & Donaldson, C. (2014). Breaking the addiction to technology adoption. *Health economics, 23*(4), 379–383. https://doi.org/10.1002/hec.v23.410.1002/hec.3034

Cooper-Jones, B., Mason, J., & Dunfield, L. (2022). Virtual care and health technology assessment considerations. *Healthcare Management Forum, 35*(3),127–9. https://doi.org/10.1177/08404704211061954

Greenhalgh, T., Jackson, C., Shaw, S., & Janamian, T. (2016). Achieving research impact through co-creation in community-based health services: literature review and case study. *The Milbank Quarterly, 94*(2), 392–429. https://doi.org/10.1111/1468-0009.12197

Holliday, N., Magee, P., & Walker-Clarke, A. (2015). Proceedings of the 3rd European Conference on Design4Health. http://research.shu.ac.uk/design4health/wpcontent/uploads/2015/07/D4H_Holliday_et_al.pdf

Juzwishin, D., McNeil, H., Ahn, J., Chen, Y., Cicchetti, A., Kume, N., Brooks, L., & Stolee, P. (2018). Aging and health technology assessment: An idea whose time has come. *Technology Assessment in Health Care, 34*(5), 442–46. https://doi.org/10.1017/S0266462318000600

Keown O.P., Parston, G.H., Patel, F., Rennie, F., Saoud H.A., & Kuwari, et al. (2014) Lessons from eight countries on diffusing innovation in health care Health Affairs, *33* (9) https://doi.org/10.1377/hlthaff.2014.0382

McNeil, H., McMurray, J., Byrne, K., Grindrod, K., McKinnon, A., & Stolee, P. T. (2017). Engaging older adults and their caregivers in innovation ecosystems for health and aging. Innovation in Aging, 1(Suppl 1), 212.

MacNeil, M., Koch, M., Kuspinar, A., Juzwishin, D., Lehoux, P., & Stolee, P. (2019). Enabling health technology innovation in Canada: Barriers and facilitators in policy and regulatory processes. *Health Policy, 123*(2), 203–14. https://doi.org/10.1016/j.healthpol.2018.09.018

Scotland, G., & Bryan, S. (2017). Why do health economists promote technology adoption rather than the search for efficiency? A proposal for a change in our approach to economic evaluation in health care. *Medical Decision Making, 37*(2), 139–147. https://doi.org/10.1177/0272989X16653397

Webster, C. (2015). E-health progress still poor $2 billion and 14 years later. *Canadian Medical Association Journal, 187*(10), E309–10. https://doi.org/10.1503/cmaj.109-5088

Concluding Remarks About the Future 11

Introduction

Through the course of this book the reader is introduced to several personas and scenarios: Mr. B and Mrs. C living in their home setting with the onset of frailties, Agnes in residential LTC, Zahra seeking access and coordination through primary care, and Francisco, a homeless and at risk older person finding himself in the interstitial space among conventional institutions of care. What all these personas demonstrate is that older adults encounter many challenges interfacing with a health care delivery system that is poorly designed to address the unmet needs of older adults. In addition, we demonstrate that current health care delivery systems are not well coordinated or integrated which makes it particularly challenging for AgeTech to be effectively mobilized to help address the challenges of older persons. In response to these limitations, we have identified a multitude of approaches and opportunities where health reforms coupled with AgeTech can make a difference in the lives of older persons (Fig. 11.1).

We identified in chapter two the primary biophysical limitations that accompany aging and the global challenges arising from chronic diseases and co-morbidities of older persons. This led us to identifying the research and innovation challenges that state health care systems will need to be prepared to respond to in the next five decades. Governments have a responsibility to mobilize long term planning to address the dramatic demographic shifts on the one hand and changing values and expectations of older persons on the other. Governing on the basis of 4-year election cycles and press releases will provide too little too late.

We described why health care delivery systems developed and evolved in siloed governance and delivery structures in response to the health needs of the population in the past. Historically supports have taken the form of pensions, income assistance, hospital

D. Juzwishin et al., *AgeTech Innovations in Healthcare for Older Adults*, Synthesis Lectures on Technology and Health, https://doi.org/10.1007/978-3-031-86885-6_11

Fig. 11.1 Embracing the future with care. *Credit* Shutterstock: PeoplesImages.com—Yuri A., https://www.shutterstock.com/image-photo/hug-relax-senior-couple-on-sofa-2352115489

and physician care, residential LTC, and to a far lesser degree health promotion or prevention, home care, primary care and rehabilitation. These supports were a necessary step in the provision of health services to the public, but they were undertaken as discrete legislative, policy, and funding interventions without taking a *continual learning systems* perspective anticipating citizens values and expectations. These systems had 'baked in' unforeseen consequences that have become the barriers to further reform today. We put forth several visionary conceptual models of an idealized health system functioning that could transform care beyond these limitations.

We demonstrated that the values of citizens generally and older persons specifically have morphed toward a need and desire to retain their independence, dignity and zest for life for as long as practically possible. Whether the setting is the home, a rehabilitation facility, long term residential care, primary care or the interstitial space where the older adults experiencing poverty dwell, we discover that the opportunity for older adults to feel fulfilled is constrained. These limitations do not arise from any malevolent intent by those delivering the service; rather, they reflect a significant lack of imagination in co-creating, legislating, and allocating funds or human resources to develop policies and programs that address emerging values and unmet needs. The tension between the old and new values and their expression have created the ethical conundrums which we identify as needing a societal response on how our aging persons ought to be supported. AgeTech should support older persons to remain in the residence of choice—to age in place for as long as possible.

Addressing System Dysfunctionality Through Reform Through Research and Innovation

Technological innovation (AgeTech) cannot compensate for inadequate or inappropriate resource availability such as deficiencies in staffing, knowledge, understanding, skills and data. Technological innovation demands greater awareness of the need for education in knowledge, skills and lifelong learning for both providers, policymakers and older persons.

We have attempted in each of the chapters to identify the necessary resourcing, education and training of health care leaders, providers, researchers, innovators, industry and policymakers to encourage a smooth life trajectory of older persons. The solutions are not only the addition of more funds as in the case of increasing the hourly per diem rates supporting residents in LTC but also examining and introducing imaginative life enhancing experiences and activities with recreation therapists and other forms of human expression in the arts and crafts. We encourage research and innovation by partnering and co-developing with older adults seeking to understand what their needs are and listening to them as to how the gaps in service could be filled.

The Tragedy of the Siloed Health Service Mentality

Agents comprising the health care delivery systems are prescribed their governance and administrative roles and responsibilities by the legislation, regulations, vision, mission, goals and objectives of their approved programs which have funds allocated to them to deliver the services. The rules surrounding the accountability, reporting and audits surrounding the expenditure of funds is monitored and closely scrutinized by the jurisdictions providing the funds. The strength of this rigid accountability is also its weakness. It diminishes the opportunity for innovative solutions with access to discretionary funds to serve the aging adult population by considering the other needs that are outside of the agent's responsibility.

AgeTech is stymied in many publicly funded health care delivery systems because of a lack of funding for what may be a risk-taking expenditure to innovate. The gap between what older persons desire to support them in a life of dignity and zest and what they are provided comprise the *ethical imperatives* that call out for imaginative health care reforms coupled with AgeTech to remedy the situation. Solutions to this challenge may be to provide some capped amount of discretionary funding for testing and developing AgeTech through streamlined regulatory processes and procurement mechanisms.

To fill this gap MEDTEQ+ and AGEWELL created the envisAGE initiative. This large-scale pan-Canadian initiative brings stakeholders together to provide resources and support to small and medium-sized enterprises to integrate technologies that address the challenges of aging and build Canadian leadership in the AgeTech market. envisAGE is funded through the Government of Canada's Strategic Innovation Fund, with additional contributions from the Government of Quebec, industry, non-profits and other partners. www.envis-age.ca.

Another approach is to explore the development of administrative tools that permit the identification of savings in one area that would be used to fund programing in another area.

A further area showing promise is the negotiation of terms of conditions among the funder, provider and industry in shared risk agreements that incentivize innovation through the identification of explicit end points that are either rewarded or penalized based on performance.

Lessons of the COVID-19 Pandemic

COVID-19, during the first two waves (March 2020 to June 2021) of the pandemic, shone a glaring light on the inadequacies of how we serve our older adults in society. Over 20,000 older Canadians died of SARS-Cov-2 in residential care facilities during that period, one of the highest mortality rates in OECD peer group comparisons. The pandemic highlighted how current health care delivery service in their current configuration are no longer fit for purpose and require reform to accommodate contemporary values of older persons living with dignity and zest. It is necessary to think about the health care delivery system as not comprised of independent elements but is an integrated system that recognizes a 'life trajectory' approach that recognizes the transitions as people age. The expectation of older persons is that their need for smooth transitions between and among the different types and levels of care and support should be provided without artificial barriers requiring them to repeat their medical history each time they are admitted to a different service or facility, or they must hand courier a copy of their immunization record between service providers. The means for an integrated and coordinated health care record considering the breadth and depth of health care interventions and supports is possible with AgeTech, what is necessary is the political will to modify legislation, regulations, policy, funding and practices to facilitate what is in the older person's interest and convenience rather than that of the provider or institution.

There is a strong expectation on the part of the older person to take responsibility for thinking about life trajectories as well. They must be realistic and prepared to address issues of anticipating what their future needs might be and planning for how they may achieve them. Lifelong learning, prudent financial planning, planning for realistic life scenarios, collaborating with researchers and innovators to co-create customized AgeTech to ensure reliability and affordability are important contributions that older adults can make. AgeTech will only be successful in its spread and scale if older adults demonstrate with confidence that it improves their ability to live with dignity and zest.

AgeTech a Global Phenomenon

AgeTech is identified as an important and necessary ingredient to improve the quality-of-life of older persons. AgeTech and its development and diffusion are global phenomena providing an opportunity for a global response. Although the UN Decade of Healthy Ageing has brought global attention to the issues most initiatives are primarily national or regional resulting in duplication of effort and false starts that are not an effective use of resources. This creates an opportunity for collaborative efforts to develop technologies jointly on a global basis to achieve rapid innovation, economies of scale and rapid innovation. Assistive technology (AT) day is a global programmatic approach on how unmet needs of older persons could be identified and prioritized to tackle through a multi-country collaborative and coordinated effort to minimize duplication and build on local expertise with a view to achieving a scaling up of the effectiveness of the diffusion of validated and effective AgeTech.

AgeTech Is Not a 'Silver Bullet' But It Can Be a Catalyst for Reform

We are under no illusion that that AgeTech is a panacea, it is not a silver bullet, it alone cannot address the challenges facing aging with dignity and zest. AgeTech should augment, but not replace, in person care giving and support. AgeTech is necessary for improving the lives of older adults, but it is not sufficient. We clearly identify and outline in each of the chapters that fundamental legislative, regulatory, policy, funding and program delivery reforms are necessary and coupled with AgeTech to achieve the improvements that are necessary. It is important that considerations of future investments in AgeTech consider that developments in one sector need to be in the context of the broader health care delivery system. Systems thinking is essential. Opportunity costs arise if spending in one sector takes away expenditures in another program that may deliver greater benefits for the community. Investment in LTC residential care should not come at the expense of improving home care or primary care supports those older adults are requesting.

AgeTech and Self-care

Digital technological innovations coupled with other innovations in care delivery create important new opportunities to support self-care by older adults. The COVID-19 pandemic highlighted the need for governments, industry and citizens to address the digital divide that serves as a barrier to access to health care services. Teleconsultation and telemedicine were the only available means of communication between citizens and health care providers during the pandemic. Virtual meetings and visits with families were the only means by which older adults could stay in contact with friends, neighbors and relatives during the pandemic. The disruptive nature of the pandemic provides a strong motivation and rationale as to why all citizens regardless of their means, geographic location, or social class ought to have access to the Internet. The pandemic also underlined the importance and necessity of supporting self-care by providing and coordinating data and information exchange in the form of an accessible and interoperable health record to both older adults and their health care providers to inform clinical decision-making and to improve the functioning of the health care system and reduce harm to individuals. It will be important for citizens, policymakers, providers, researchers and innovators to not slide back to the practices of the past.

Co-creation of AgeTech

Aging persons are heterogeneous in their characteristics therefore the concept of 'one size fits all' does not work for older persons and their relationship with AgeTech. It is essential to ensure that the health care delivery systems coupled with innovators and industry are prepared to engage with and empower older adults to participate and help inform the development, testing and implementation of new and promising AgeTech. The engagement of older persons must be culturally competent, respectful and approached with a sense of humility. Equity of access must consider that not all aging persons will be able to pay for the AgeTech that could enhance their dignity and zest for life. Legislation, policy and creative funding models need to be developed to bridge the digital divide that prevents those without the financial means to gain access to necessary AgeTech.

Older persons, their care givers, health care providers, as well as operators and administrators need to create an environment in which they are engaged in all stages of technology development and implementation. Co-creation needs to engage participation that is reflective of diverse backgrounds, cultures and health conditions. Usability must be a mandatory partnered effort of the older person, innovator and provider. Innovation needs to focus on the areas of greatest need and concern to those involved, not the easiest problems to solve. Relationships and trust among the team are critical to successful engagement. The ethical issues of privacy and confidentiality are more easily understood and solutions tested if all parties to the technology development are involved.

The emergence of best practices in how best to expedite co-design and co-creation through a collaborative approach to identify and prioritize unmet needs, build relationships on a foundation of trust, increase the awareness of opportunities, and manage all participants' expectations are all necessary ingredients for establishing and practicing best practices. The heterogeneity in the characteristics of older persons necessitates that the necessary knowledge and skills of co-creation and usability testing are continually improved upon.

Co-designing AgeTech

There is widespread recognition of the importance of user engagement and working collaboratively with all interested parties to facilitate the development and implementation of technologies or other innovations (Blomkamp, 2018; Boyd et al., 2012; McNeil et al., 2016). Bate and Robert (2006) have described the value of a shift from designing the system around the patient to co-designing services with the patient, resulting in an improved and more effective user experience. The benefits of co-design processes have been identified for a variety of technologies in a variety of health care settings for older adults (Carroll et al., 2021).

The McMaster Health Forum identified engaging LTC residents, caregivers and operators in the development and adoption of technologies as a key element of an approach to more fully realize the potential of technologies in that context (Gauvin et al., 2021) and as part of reimagining and reforming the LTC sector (Wilson et al., 2022). An important reason for active engagement of older persons throughout the design process is to ensure that the priorities and needs of older persons remain central. As has been found in the work of Wang et al. (2019) in co-designing technologies to support aging-in-place, older adults have shown a willingness and desire to participate in technology design, and feel they provide value in sharing their lived experiences and providing their own unique perspectives.

While clearly beneficial, co-design processes with older adults have also encountered challenges. In co-designing home care reminder systems with older users, McGee-Lemon et al. (2012) found a challenge related to the limited knowledge of users on the capabilities of currently available technology. Harrington and colleagues (2018) encountered similar challenges in co-designing health and fitness apps with older adults. Several co-design groups have highlighted the value of longer-term processes of engagement to give older adult users more time to interact with the technology (Harrington et al., 2018; Ostrowski et al., 2021). While valuable, longer periods of engagement could add significantly to the costs of these efforts.

Descriptions of the design processes for social robots for older people (Peine & Moors, 2015; Peine & Neven, 2021) illustrate how aging-associated deficits that could be more readily addressed by the robots (such as forgetfulness) were prioritized over aspects of aging that may have been of greater concern to older persons themselves (such as lack of social connectedness). Ostrowski and colleagues (2021) also used a co-design method to develop social robots, working with older adults in LTC. Lessons learned from their work include the importance of developing relationships with participants and creating a safe space for participation; the benefits of giving older adults time to interact with the social robot in its intended context so that users can generate a mental model of their desired robot; and the need for ongoing communication, especially if the co-design process is of long duration (Ostrowski et al., 2021).

As has been found in co-design projects in industry (Kleinsmann & Valkenberg, 2008), co-design processes with older adults have illustrated the importance of engagement of stakeholders at multiple levels. In developing virtual reality exergames for persons living with dementia, Muñoz et al. (2022) involved older adults with dementia living in LTC, older adults living in the community, exercise therapists, virtual reality game specialists and researchers. Their work also illustrated the challenges of reconciling different viewpoints and interests.

Despite its importance, co-design is often not done well (Bogza, Laura-Mihaela et.al., 2020; Sumner et al., 2021). Sumner and colleagues (2021) undertook a systematic review on the effects and experiences of co-designed technology to support aging-in-place. Their review found great variation in the extent of older adult involvement in these processes. The authors recommended that efforts to improve co-design processes should address: relationship and trust building among stakeholders; increasing knowledge among stakeholders to increase awareness of differing perspectives and to avoid unrealistic expectations; and skills and knowledge in co-design methods (Sumner et al., 2021). Involving older people living with dementia in technology development can be challenging (Suijkerbuijk et al., 2019), but has been found to provide valuable input into the design process and to have benefits for the person with dementia (Span et al., 2013; Marradi et al., 2019). Based on their systematic review of the literature, Rai and colleagues (2020) have proposed best practice guidelines for involving persons with dementia in developing technology. The guidelines emphasize careful preparation for the involvement, use of appropriate language to promote understanding, opportunities for learning new skills, involvement throughout the process, and keeping participants informed after their involvement.

The best practice guidelines proposed by Rai, Harleen Kaur; and colleagues (2020) for persons living with dementia are broadly consistent with the framework developed by McNeil and colleagues (2016) for engagement of older adults in health care research and planning. McNeil and colleague's (2016) framework proposed five stages of engagement (1) understanding the environment and context; (2) careful planning for meaningful engagement; (3) establishing relationships of trust among those involved; (4) building a sustainable relationship; and (5) transitioning the engagement as the project closes—the transition phase involves providing timely feedback and information that is sensitive to culture and literacy level.

Ageism

The marginalization of older adults, particularly those who are at risk, often results in a commensurate deprecation in the value of those who are commissioned or expected to provide support to them. Personal service workers in LTC residential facilities are often entry level positions often left to immigrant workers that are employed because of the relatively low-cost labor that they provide. The marginalization of these workers is often associated with a minimal level of support for technical upgrading of their skills and knowledge. This was particularly evident during the recent pandemic when knowledge of infection control protocols and use of protective equipment by the PSWs and other staff was insufficient to address the urgency the situation demanded. Adoption and implementation of technologies must be supported by training and other education opportunities to support the use, implementation, and integration of the AgeTech into workflows and existing practices and ongoing maintenance.

Another example of the marginalization by association is the value that is afforded toward family members, relatives and other volunteers who support aging persons. They are often not valued for the contribution that they make to support older adults. Future investments in AgeTech will need to consider these staff and volunteers to ensure that future investments in innovations benefit them as well.

Concluding Remarks

The reader has had a broad exposure to the challenges and issues associated with identifying ways to improve health care delivery systems coupled with AgeTech. We are only beginning to scratch the surface of the possibilities afforded by the digital technology revolution that we are witnessing. We are convinced that AgeTech devoid of human contact expressing empathy and care is certain to fail. Novel approaches and yet unproven models for delivery across the continuum of care are emerging. Additional research is needed on the human factors, feasibility, usability, and effectiveness of technological innovations, including their cost effectiveness. Toward that end projects are identified in each of the chapters as a challenge to those in the health care delivery systems to improve the opportunity for older persons to enjoy a life with dignity and zest.

Readers are encouraged to engage in discussions with their peers of the questions that are posed in Appendix A to stimulate further developments in the AgeTech space.

References

Blomkamp, E. (2018). The promise of ci-design for public policy. *Australian Journal of Public Administration, 77*(4), 729–43. https://doi.org/10.1111/1467-8500.12310

Boyd, H., McKernon, S., Mullin, B., & Old, A. (2012). Improving healthcare through the use of co-design. *The New Zealand Medical Journal, 125*(1357), 76–87. PMID: 22854362.

Bogza, L. M., Patry-Lebeau, C., Farmanova, E., Witteman, H. O., Elliott, J., Stolee, P., & Giguere, A. M. (2020). User-centered design and evaluation of a web-based decision aid for older adults living with mild cognitive impairment and their health care providers: mixed methods study. *Journal of medical Internet research, 22*(8), e17406. https://doi.org/10.2196/17406

Carroll, S., Kobayashi, K., Cervantes, M. N., Freeman, S., Saini, M., & Tracey, S. (2021). Supporting healthy aging through the scale-up, spread, and sustainability of assistive technology implementation: A rapid realist review of participatory co-design for assistive technology with older adults *Gerontology and Geriatric Medicine, 7*. https://doi.org/10.1177/23337214211023269

Harrington, C. N., Wilcox, L., Connelly, K., Rogers, W., & Sanford, J. (2018). Designing health and fitness apps with older adults: Examining the value of experience-based co-design. *In Proceedings of the 12th EAI international conference on pervasive computing technologies for healthcare* 15–24.

Kleinsmann, M., & Valkenburg, R. (2008). Barriers and enablers for creating shared understanding in co-design projects. *Design studies, 29*(4), 369–386. https://doi.org/10.1016/j.destud.2008.03.003

Marradi, C., Wang, G., Albayrak, A., & van der Cammen, T. J. (2019). Co-designing with people with dementia: A scoping review of involving people with dementia in design research. *Maturitas, 127*, 55–63. https://doi.org/10.1016/j.maturitas.2019.06.003

Muñoz, J., Mehrabi, S., Li, Y., Basharat, A., Middleton, L. E., Cao, S., & Boger, J. (2022). Immersive virtual reality exergames for persons living with dementia: user-centered design study as a multistakeholder team during the COVID-19 pandemic. *JMIR Serious Games, 10*(1) e29987 https://doi.org/10.2196/29987

McGee-Lennon, M., Smeaton, A., & Brewster, S. (2012). Designing home care reminder systems: lessons learned through co-design with older users. In 2012 6th International Conference on Pervasive Computing Technologies for Healthcare (PervasiveHealth) and Workshops 49–56.

McNeil, H., Elliott, J., Huson, K., Ashbourne, J., Heckman, G., Walker, J., & Stolee, P. (2016). Engaging older adults in healthcare research and planning: A realist synthesis. *Research Involvement and Engagement, 2*, Article no. 10. https://doi.org/10.1186/s40900-016-0022-2

McMaster University Health Forum, Identifying and Harnessing the Potential of Technology in Long-term Care Settings in Canada https://www.mcmasterforum.org/docs/default-source/product-documents/evidence-briefs/tech-in-ltc-eb.pdf?sfvrsn=9bcb0fb_4&utm_source=chatgpt.com

Ostrowski, A. K., Harrington, C. N., Breazeal, C., & Park, H. W. (2021). Personal narratives in technology design: the value of sharing older adults' stories in the design of social robots. *Frontiers in robotics and aI, 8*, 716581–716581. https://doi.org/10.3389/frobt.2021.716581

Paul., Bate G. (2006). *Robert bringing user experience to healthcare improvement the concepts methods and practices of experience-based design*CRC press London

Peine, A., & Moors, E. H. (2015). Valuing health technology–habilitating and prosthetic strategies in personal health systems. *Technological Forecasting and Social Change, 93*, 68–81. https://doi.org/10.1016/j.techfore.2014.08.019

Peine, A., & Neven, L. (2019). From intervention to co-constitution: New directions in theorizing about aging and technology. *The Gerontologist, 59*(1), 15–21. https://doi.org/10.1093/geront/gny050

Rai, H. K., Cavalcanti Barroso, A., Yates, L., Schneider, J., & Orrell, M. (2020). Involvement of people with dementia in the development of technology-based interventions: narrative synthesis review and best practice guidelines. *Journal of medical Internet research, 22*(12), e17531. https://doi.org/10.2196/17531

Sumner, J., Chong, L. S., Bundele, A., & Wei Lim, Y. (2021). Co-designing technology for aging in place: A systematic review. *The Gerontologist, 61*(7), 395–409. https://doi.org/10.1093/geront/gnaa064

Suijkerbuijk, S., Nap, H. H., Cornelisse, L., IJsselsteijn, W. A., De Kort, Y. A., & Minkman, M. M. (2019). Active involvement of people with dementia: A systematic review of studies developing supportive technologies *Journal of Alzheimer's Disease 69*(4), 1041–1065. https://doi.org/10.3233/JAD-190050

Span, M., Hettinga, M., Vernooij-Dassen, M., Eefsting, J., & Smits, C. (2013). Involving people with dementia in the development of supportive IT applications: a systematic review. *Ageing research reviews, 12*(2), 535–551. https://doi.org/10.1016/j.arr.2013.01.002

Wilson, M. G., Gauvin, F. P., DeMaio, P., Alam, S., Drakos, A., Soueidan, S., & Lavis, J. N. (2022). Enhancing the use of technology in the long-term care sector in Canada: Insights from citizen panels and a national stakeholder dialogue. *35*(5), 310–317. https://doi.org/10.1177/08404704221108466

Wang, S., Bolling, K., Mao, W., Reichstadt, J., Jeste, D., Kim, H. C., & Nebeker, C. (2019). Technology to support aging in place: Older adults' perspectives healthcare *7*(2), 60. https://doi.org/10.3390/healthcare7020060

Questions and Challenges for Readers

A

This section of the book is intended to inspire readers whether they are older adults, students, educators, providers, policymakers, business owners, innovators or leaders to think about stretching outside of the prescribed boundaries and path dependency of contemporary health care services and delivery and imagine a world where older persons can live a fulfilled life of dignity and zest.

AgeTech can only be successful as a partner to reform if all the participants of the health care system are supportive of co-creating with older persons incentives and opportunities for AgeTech to be successful.

Thinking through the issues presented and pressing yourself to be observant of the practical challenges and ethical issues outlined earlier in the book will encourage you to think, plan and act from a systems continual learning perspective. The reader is encouraged to carry out thought experiments and engage in challenge discourses with older persons and colleagues to situate the problems and issues identified below within the context of a modernized legislative, regulatory, policy, funding and practice landscape. In this way undertakings to develop and apply AgeTech to cross the digital divide are less likely to be curtailed by barriers that make diffusion difficult.

- How will the two trends of the increasing life span among men and a dramatic growth of the 85 + population impact on living arrangements, community and family structures in your community? How can AgeTech support these emerging conditions? What legislative, policy, funding and regulatory reforms are necessary to address the emerging needs?
- The incidence of chronic conditions has been identified as more important than age in determining disability and health care resource consumption. How can AgeTech be mobilized to stimulate health promotion and disease prevention to help improve the quality-of-life of the older adults experiencing poverty and ease the demand on

D. Juzwishin et al., *AgeTech Innovations in Healthcare for Older Adults*, Synthesis Lectures on Technology and Health, https://doi.org/10.1007/978-3-031-86885-6

health care delivery utilization? What legislative, regulatory, policy, funding, or practice reforms are necessary? How and who would you approach to address these?

- Identify an older person in your life who may be at risk but exhibits characteristics of positive aging. What do they do that contributes to their positive approach to life and aging? Does AgeTech play a role in their life trajectory? How does it nurture a positive approach? How in a negative way? How can the legislative, funding and policy landscapes be shaped to nudge positive aging?
- How have social media contributed to the diminishment of power and the increase of ageism in the at risk older adults? How might AgeTech be used to counter these phenomena? How would you respond to someone making ageist remarks? What reforms in the use and exploitation of social media are necessary to preserve democratic values, respectful exchange, and civil society? How can AgeTech help? How is AgeTech not helpful? How do we address that?
- What issues are emerging in respect to the application of generative artificial intelligence to the AgeTech industry and aging? How would you critically respond to the emerging issues and challenges? How would you work toward making a positive contribution?
- What are the positive contributions that generative artificial intelligence can make to the development, application and evaluation of AgeTech? How would you go about strengthening those contributions?
- Provide examples of where and how AgeTech has empowered the at-risk older persons. Why have those been successful? Are there examples of where AgeTech has diminished the power of the at-risk older persons? How can we respond to those influences? What do we need to do to repair this?
- How will you prepare yourself for the 100 year life cycle? What kinds of legislation, regulations, policy and funding or insurance programs are necessary to deal with this demographic shift? How can AgeTech be a help in this shift?
- The Decade of Healthy Ageing provides an opportunity to explore and forge new relationships within and outside jurisdictions to act on healthy aging. What partnerships on the global scale can be forged to stimulate the effective mobilization of AgeTech to contribute? What contribution will you make toward the decade of healthy aging? How will go about operationalizing your ideas?
- How can AgeTech be mobilized to create and stimulate a more positive narrative within our health care system of our aging society particularly those who are at risk? What institutional initiatives coupled with AgeTech could have a positive influence? What would be some of the ingredients for a positive narrative? How would you validate the narrative? How will you disseminate the narrative? How would you implement the narrative?
- How can we communicate and influence law and health policy that is important and necessary to invest in legislation, regulations, policies and practices that promote good

health in aging and well-being? Is there a role for AgeTech in this space? What might the role be and how would you assess whether it is successful?

- How would you go about inviting an aging adult living in poverty to participate in your research or AgeTech innovation project about improving access to health and social care? How would you situate them and their voice in the group discussions? How would you gain their trust and respect? How would you acknowledge their contributions?
- Think of the many different forms and conditions in which the older adults experience poverty in your community. What three AgeTech undertakings coupled with reforms could contribute to making their shelter and access to employment more friendly? How would you test the validity of your assumptions?
- Identify and describe three housing models that are available in your community. How can AgeTech be mobilized to improve the smooth access, usability and comfort of the three models in its relationship to health care delivery to the aging adult experiencing poverty? Who would you work with to advance your ideas?
- What are three major transportation challenges for the at-risk older adult and their access to health and social care? What institutional or funding reforms are necessary to expedite access? How can AgeTech be mobilized to respond to these? What unforeseen circumstances might arise?
- Address the issue of whether AgeTech means an inevitable loss of privacy and dignity for the aging adult living in poverty? Describe why these concepts are not mutually exclusive and how one can achieve both? What characteristics of AgeTech must be demonstrated to achieve both privacy and ubiquity of health information when it is called for?
- Identify the forms of harm that come to older persons because of a lack of interoperability of health care data in the health care system. How would you orchestrate reforms to address these issues?
- AgeTech has both strengths and weaknesses. Identify three common AgeTech interventions that do not work well for the at-risk older adult. Describe their strengths and weaknesses. How can the weaknesses be mitigated? How would you go about testing the usability of the refinements?
- AgeTech that is oriented toward medical interventions often overshadows AgeTech for health promotion or disease prevention that responds to the challenges of frailty in older adults. Why is that so? How would you organize and incentivize health care funding, providers and system reform to better address issues of frailty in older adults?
- Does the application of AgeTech in the health care delivery system magnify or diminish or devalue human interaction and the character of caring and empathy? Provide examples of interactions that diminish dignity. What interactions promote dignity? How can this be mitigated through education and behavior change coupled with reforms?

- Does the integration of generative artificial intelligence capacity in AgeTech add value or does it devalue human interaction? What steps can be taken to increase value and prevent a devaluing? How would you know?
- Provide three examples of AgeTech in health care delivery where the product is a marketing gimmick aiming to exploit at-risk older people. How would you address this exploitation? Would you have the temerity to challenge the exploitation? What institutional or regulatory changes are necessary to avoid this exploitation? How can they be mobilized?
- Policymakers and administrators in health and social care delivery systems rarely see the addition of new devices, such as AgeTech as being a saving or cost avoidance for a sustainable health care system. How would you make and deliver a pitch for the introduction of a novel and theoretically effective AgeTech device to an assistive device funder? Are there institutional barriers that must be accounted for? How would you go about addressing them?
- How would you ensure that the conceptualization, protype development and testing of your AgeTech device would be turned into a real world product that funders and at-risk older persons would be wanting to adopt? Are their barriers or disincentives in the health care delivery system that you need to be aware of? How would you address them?
- Explain how social isolation and loneliness are interrelated but different. What AgeTech solutions may be appropriate for addressing each of these? What legislative, regulatory, policy, funding or institutional issues need to be addressed to facilitate AgeTech success? What AgeTech solutions might increase social isolation and loneliness? How would you address those?
- What five barriers hold the at-risk older persons from accessing the health care delivery system? What institutional barriers stand in the way? How could AgeTech help remove those barriers? How would you mobilize the facilitators of access?
- You have been asked to participate in a procurement specification development for acquiring AgeTech to address issues of at risk older persons. Are there criteria and deliverables in the procurement specifications that conflict with bringing dignity and respect to the at-risk older persons? What might they be? How would you address them? What are your recommendations for undertaking the tender?
- You have been asked to participate in the framing of a grant application process to invite research into AgeTech to increase the autonomy and dignity of older persons. What requirements would you request applicants to provide? How would you assess and adjudicate on the submitted grants?
- Pets can be an important part of people's lives; with the at risk older adult this can be particularly important. How would you get a glimpse into the importance of pets in the lives of at risk older adults? What conveniences or utilities could AgeTech play in the life an at-risk person that would enhance their lifestyle with a pet? Are there

institutional barriers that must be addressed for the AgeTech to be successful? How would you go about addressing those barriers?

- Therapy animals can be important companions to the at risk older persons. What are some AgeTech innovations that would enhance the lives of the therapy animal and the individual? Are there societal barriers to the use of therapy animals? How would you address them?
- You have been asked to participate in the design of a health promotion educational app for the at-risk older persons on sexuality and health. What are the strengths and weaknesses of current apps in use? How would go about addressing them? What are your six key messages?
- You have been asked to participate in the development of an AgeTech device and app to counter elder abuse. How would you gain an understanding of the issues facing victims of elder abuse? Describe activities you would suggest for identification, detection, monitoring, reporting, prevention or direct intervention. What issues would you need to navigate? How and who would you consult and involve?
- Mandatory reporting of elder abuse has potential risks and benefits. Often risk and benefits are conflated or mutually exclusive. Deconstruct and map the ethical dimensions of the risks and benefits. Can AgeTech coupled with institutional, or policy solutions help address the challenges associated with those risks and benefits? How?
- Self-reporting of mistreatment in older adults is low. What are the societal barriers? What are the cultural barriers? What are the behavioral barriers? What can be done to help improve self-reporting? Could AgeTech be instrumental in aiding this challenge? How?
- How can the philosophies and wellness models of Indigenous peoples influence health promotion activities to enhance the lives of older Indigenous peoples? Is there a role for AgeTech? How would you know? How would you validate your assumptions?
- What are some key considerations in developing effective health services access to the immigrant communities? Can AgeTech help bridge barriers to access? How and who would you approach to address this challenge?
- What are some of the lifelong learning needs of at risk older adults? How could societal interventions coupled with AgeTech facilitate opportunities for learning and how would you encourage them? What barriers would you need to address? How would you address them? How would you measure the success of your idea?
- How are lifelong learning opportunities in your province or community nurtured? Is there a role for AgeTech to help the at risk older adult access and maintain lifelong learning? How would you do it?
- What programs promote health literacy in your community? Explain why health literacy is important for health promotion of the at risk older adult? What are some barriers? How might health literacy be encouraged and facilitated with AgeTech?

- What has the COVID-19 pandemic taught us about its impact on the aging older person experiencing poverty? What institutional and policy barriers need to be addressed? How can AgeTech be used to mitigate the negative impact of a future epidemic?
- Climate change has created a volatility in the weather patterns around the world. What are the serious manifestations of these patterns that will have the greatest negative impact on older persons? How should those impacts be identified and addressed? How would you mobilize plans and actions to respond to these challenges?
- Nutrition is key to a healthy lifestyle. How well do we understand the data and information about nutrition of the at risk older adults in our community? What are some of the barriers to their understanding? How can AgeTech be used to provide support to the at risk older adults in maintaining a healthy diet?
- How did COVID-19 impact on the at risk older adult's dietary intake? What data and information do we have? Is there a role for AgeTech to ameliorate those challenges? How would you do it?
- Staying physically active is an important contributor to a healthy lifestyle. What institutional factors stifle physicality of older persons? How can they be addressed? Can AgeTech provide support to the at-risk older persons in maintaining their physical capacity? How would you do that?
- In what way do risk and safety come into conflict for the at-risk older persons? In what ways are they mutually supportive? Is there a role for AgeTech to mitigate between the two tensions to ameliorate them for the at risk older adult? Who would be interested in your ideas as to how to mitigate risk?
- Describe some promising features of suicide prevention programs for older persons. How are these different from MAID? How are the conditions or circumstances different for the at risk older adult? Is there a role for AgeTech to mitigate the risk of suicide? How would you go about doing a risk mitigation strategy?
- Unpaid care givers and volunteers play an important role in providing support to the at-risk older adults in society. Cash for care has been demonstrated to be effective in supporting carers. How might AgeTech be used to support a community of care givers and volunteers to reach and support the at-risk older persons more effectively?
- A non-profit organization has requested that you help them develop a program accompanied by an app to support the at risk older adults in your community. Are there institutional or policy barriers that need to be addressed? What steps would you go through to respond to the barriers and develop this idea?
- You have been requested to develop a program to provide smart devices to the at-risk older adults in your community. How would you determine what the older adults want or need? What steps would you go through to develop the policy, practice, funding, distribution and monitoring of the smart devices? How will you know if you are successful?
- Some older adults experiencing poverty might be reluctant to seek services for substance abuse. How would you work with the older adults to gain their trust and identify

the barriers to service access and explore whether an AgeTech solution would address the older adults' concerns?

- What are the experiences and components of a good death of individuals that are at-risk older adults in your community? Are their issues different from those of other older persons? Are institutional or practice barriers in the way of a good death? How would you know what a good death for an at risk older adult is? How might AgeTech be mobilized to deliver palliative care or hospice care?

Glossary

A

Accessibility "The degree of ease that something (for example, device, service, physical environment and information) can be accessed, used and enjoyed by persons with disabilities. The term implies conscious planning, design and/or effort to make sure something is barrier-free to persons with disabilities. Accessibility also benefits the general population, by making things more usable and practical for everyone, including older people and families with small children…." *Source* Accessibility in Ontario, Development of proposed postsecondary education standards—final recommendations report 2022. Appendix B: Glossary of terms and definitions. https://www.ontario.ca/page/development-proposed-postsecondary-education-standards-final-recommendations-report-2022#section-13

AgeTech as any digital technology that aims to disrupt, extend, expand or re-organize existing social practices, leading to new, more agile opportunities to enhance independent living of older adults in comfort and security

Aging-in-place "having the health and social supports and services you need to live safely and independently in your home or your community for as long as you wish and are able" *Source* AgeCare. (n.d.). https://www.agecare.ca/aging-place-really-means/

Arrythmia "is an abnormal heart rhythm for you, but not necessarily for someone else" *Source* Heart & Stroke (2023). https://www.heartandstroke.ca/heart-disease/conditions/arrhythmia

Artificial intelligence "It is the science and engineering of making intelligent machines, especially intelligent computer programs. It is related to the similar task of using computers to understand human intelligence, but AI does not have to confine itself to methods that are biologically observable."

"At its simplest form, artificial intelligence is a field, which combines computer science and robust datasets, to enable problem-solving. It also encompasses sub-fields of machine learning and deep learning, which are frequently mentioned in conjunction

D. Juzwishin et al., *AgeTech Innovations in Healthcare for Older Adults*, Synthesis Lectures on Technology and Health, https://doi.org/10.1007/978-3-031-86885-6

with artificial intelligence. These disciplines are comprised of AI algorithms which seek to create expert systems which make predictions or classifications based on input data."

Source What is artificial intelligence? IBM. https://www.ibm.com/topics/artificial-intelligence

Assisted living "is for people who need help with daily care, but not as much help as a nursing home provides. Assisted living facilities range in size from as few as 25 residents to 120 or more. Typically, a few 'levels of care' are offered, with residents paying more for higher levels of care. Assisted living residents usually live in their own apartments or rooms and share common areas. They have access to many services"

Source National Institute on Aging (n.d.). Residential facilities, assisted living, and nursing homes. https://www.nia.nih.gov/health/residential-facilities-assisted-living-and-nursing-homes

Autonomy "Personal autonomy is, at minimum, self-rule that is free from both controlling interference by others and from limitations, such as inadequate understanding, that prevent meaningful choice. The autonomous individual acts freely in accordance with a self-chosen plan, analogous to the way an independent government manages its territories and sets its policies."

Source Varelius, J. (2006). The value of autonomy in medical ethics. *Medicine, Health Care and Philosophy*, 9(3), 377–388. https://doi.org/10.1007/s11019-006-9000-z

B

Biological age "refers to epigenetic alteration and DNA methylation which express on how able and functioning she is and whether she has diseases related to old age"; "Does not increase at the same rate for everyone"

*Source*Räsänen, J. (2021, April 1). Centre for Bioethics, Harvard Medical School, Does the distinction between biological and chronological age support legal age change? https://bioethics.hms.harvard.edu/journal/legal-age-change

Blue-ribbon reports "of outstanding quality"; "*especially*: consisting of individuals selected for quality, reputation, or authority"

Source Merriam-Webster, (2023). https://www.merriam-webster.com/dictionary/blue-ribbon

Boundary spanners "institutions, groups, or individuals that straddle the divide between information producers and users (eg scientists and non-scientists, respectively), produce boundary products or tools ("boundary objects") that enable communication between these two groups, and are accountable in some fashion to both groups"; "two central functions relate to information acquisition and delivery"

Source Safford, H. D., Sawyer, S. C., Kocher, S. D., Hiers, J. K., Cross, M. (2017, December 1). Linking knowledge to action: The role of boundary spanners in translating ecology, *Frontiers in Ecology and the Environment*, *15*(10), 560–8. https://doi.org/10.1002/fee.1731

Boundary spanning "the process of search for knowledge beyond existing boundaries such as organizational, technological, temporal or geographic"

Source Nerkar, A., & Miceli, K. A. (2016). Boundary spanning. In M. Augier & D. Teece (Eds.). *The Palgrave Encyclopedia of Strategic Management*. Palgrave Macmillan. https://doi.org/10.1057/978-1-349-94848-2_333-1

C

Capacity building "the process of developing and strengthening the skills, instincts, abilities, processes and resources that organizations and communities need to survive, adapt, and thrive in a fast-changing world"

Source United Nations (n.d.). Academic impact https://www.un.org/en/academic-impact/capacity-building#:~:text=Capacity%2Dbuilding%20is%20defined%20as,in%20a%20fast%2Dchanging%20world.

Champions "The project champion is the person within an organization implementing a project who takes on the burden of ensuring everyone involved is on board and behind the ultimate success of the project"

Source Miles, C. (2021, July 26). The project champion: A management best practice. *Small Business Trends*. https://smallbiztrends.com/2013/12/what-is-a-project-champion.html

Change agents "a person who acts individually or collectively to influence, instigate, promote, or enable a positive change for health care transformation"

Source **Accelerating Change Transformation Team,** Alberta Medical Association. (2023). https://actt.albertadoctors.org/blueprint/change-agents/

Chronological age "refers to the actual amount of time a person has existed"; "increases at the same rate for everyone"

*Source*Räsänen, J. (2021, April 1). Centre for Bioethics, Harvard Medical School, Does the distinction between biological and chronological age support legal age change? https://bioethics.hms.harvard.edu/journal/legal-age-change

Chronic diseases Government of New Brunswick (n.d.) "are prolonged conditions that generally cannot be prevented by vaccines or cured completely. Many chronic diseases develop slowly over time sometimes because of poor health behaviors such as smoking or inactivity. Diabetes, arthritis, depression, asthma, congestive heart failure are examples of chronic diseases"

Source https://www2.gnb.ca/content/gnb/en/departments/health/patientinformation/PrimaryHealthCare/What_is_Primary_Health_Care/Chronic_Disease_Prevention_and_Management.html

Clostridium difficile/clostridioides difficile "bacterium that causes mild to severe diarrhea and intestinal conditions like pseudomembranous colitis (inflammation of the colon). *C. difficile* is the most frequent cause of infectious diarrhea in hospitals and long-term care facilities in Canada"; "is a bacterium that causes an infection of the large intestine (colon). Symptoms can range from diarrhea to life-threatening damage to the colon. The bacterium is often referred to as C. difficile or C. diff."

Sources Government of Canada (2014). Fact Sheet: Clostridium difficile (C. difficile). https://www.canada.ca/en/public-health/services/infectious-diseases/fact-sheet clostridium-difficile-difficile.html; Mayo Clinic (2023). C. difficile infection. https://www.mayoclinic.org/diseases-conditions/c-difficile/symptoms-causes/syc-20351691

Cognition capacity "Dr. Olenka Bilash, at the University of Alberta, defines cognitive capacity as the total amount of information the brain is capable of retaining at any particular moment.' She gives the example of attention and points out that different activities take up different amounts of our brain's attention capacity"

Source Hill, B., & Stark, R. (n.d.). "What do you mean by 'Cognitive capacity?'", Brain Ware. https://mybrainware.com/blog/what-do-you-mean-by-cognitive-capacity/

Commonwealth Fund "We support independent research on health care issues and make grants to promote better access, improved quality, and greater efficiency in health care, particularly for society's most vulnerable"

Source The Commonwealth Fund (2023). About us. https://www.commonwealthfund.org/

Co-morbidity "a pre-existing medical condition of a patient, or the presence of one or more medical conditions known to increase risk of death, that exist in addition to the most significant condition... that causes a patient's stay in the hospital. The number of comorbid conditions can be used to provide an indication of the health status (and risk of death) of patients. In other words, comorbidity is an indicator of the differential utilization of hospital care"

Source University of Manitoba, Manitoba Centre for Health Policy (n.d.). Term: Comorbidity/Comorbidities. http://mchp-appserv.cpe.umanitoba.ca/viewDefinition.php?definitionID=102446

Co-development Co-development is an approach to research in which end users and other stakeholders actively participate in the process of innovation throughout a project. Taking an equitable approach to co-development means giving equal standing to the knowledge and expertise of all participants and providing an environment in which everyone is able to contribute. It also means recognising that the distribution of the real-world benefits of an innovation will be informed by the social, cultural, political and economic context in which the innovation is eventually used.

Stockholm Environment Institute https://www.york.ac.uk/sei/equitable-tech/

Co-production "Is an approach to decision-making and service design rather than a specific method. It stems from the recognition that if organisations are to deliver successful services, they must understand the needs of their users and engage them closely in the design and delivery of those services. Co-production rejects the traditional understanding of service users as dependents of public services, and instead redefines the service/user relationship as one of co-dependency and collaboration. Just like users need the support from public services, so service providers need the insights and expertise of its users in order to make the right decisions and build effective services."

Source Involve (n.d.). Co-production. https://www.involve.org.uk/resource/co-production

D

Data capture "is the process of extracting information from any type of structured or unstructured document (paper or electronic) to transform it into a machine-readable digital format"

Source Jameson, S. (2021). Data capture: Definition, process, methods, and benefits. Nanonets. https://nanonets.com/blog/what-is-data-capture/

Differentiation "development from the one to the many, the simple to the complex, or the homogeneous to the heterogeneous" and **"**the act or process of differentiating"

Source Merriam Webster (n.d.). differentiation. https://www.merriam-webster.com/dictionary/differentiation#:~:text=noun,the%20homogeneous%20to%20the%20heterogeneous

Digital learning "is any type of learning that uses technology. It can happen across all curriculum learning areas"

Source Victoria State Government, Australia. Department of Education. (2019). Teach with digital technologies. https://www.education.vic.gov.au/school/teachers/teachingresources/digital/Pages/teach.aspx#:~:text=Digital%20technologies%20are%20electronic%20tools,across%20all%20curriculum%20learning%20areas.

Digital technologies "are electronic tools, systems, devices and resources that generate, store or process data. Well known examples include social media, online games, multimedia and mobile phones"

Source Victoria State Government, Australia. Department of Education. (2019). Teach with digital technologies. https://www.education.vic.gov.au/school/teachers/teachingresources/digital/Pages/teach.aspx#:~:text=Digital%20technologies%20are%20electronic%20tools,across%20all%20curriculum%20learning%20areas.

E

Empowerment "Refers to increasing the power of a low power group so that it is equal to the high power group. Empowerment is the interpersonal process of providing the proper tools, resources, and environment to build, develop, and increase the ability and effectiveness of others to set and reach individual goals." "Is an intrinsic motivation made up of four cognitions: meaning, competence, self-determination, and impact. it is a process by which people, organizations, and communities gain power."; "Empowerment reflects a quality or property a nurse possesses, allowing for influence over their environment. The focus of empowerment as an outcome is more on solutions than problems."; "At the community level is people uniting to accomplish common goals. Two underlying assumptions ... First, every person has the potential to become empowered. Second, it is an assumption that empowered communities develop from empowered individuals working together to achieve a mutual goal for that community."

Source Haddad, L. M., Toney-Butler, T. J. (2023). Empowerment. National Library of Medicine. https://www.ncbi.nlm.nih.gov/books/NBK430929/

***Escherichia coli* (*E. coli*) O157:H7** "bacteria found in the environment, foods, and intestines of people and animals. *E. coli* are a large and diverse group of bacteria. Although most strains of *E. coli* are harmless, others can make you sick. Some kinds of *E. coli* can cause diarrhea, while others cause urinary tract infections, respiratory illness and pneumonia, and other illnesses"

Source Centers for Disease Control and Prevention. (2022). *E. coli (Escherichia coli)* https://www.cdc.gov/ecoli/index.html

Essential tremor "is a common movement disorder that involves a tremor (unwanted and uncontrolled shaking) in both hands and arms during action and when standing still"

Source National Institute of Neurological Disorders and Stroke (n.d.). Essential tremor. https://www.ninds.nih.gov/health-information/disorders/essential-tremor#:~:text=What%20is%20essential%20tremor%3F,voice%20and%20how%20you%20walk

Environmental stability "is a conservation concept which is the meeting of services and resources of present and future generations without affecting the health of the ecosystems that provide them"

Source Place, S. E. (2018). Animal welfare and environmental issues. In J. A. Mench (Ed.), *Advances in agricultural animal welfare, science and practice* (pp. 69–89). Woodhead Publishing. https://doi.org/10.1016/C2015-0-05920-2

e-visits "Electronic communication between patients and physicians ("e-visits") is a recent technological innovation in primary care that affords patients a low-cost alternative to physician office visits"; "an online visit with a primary care provider that can be used as an alternative to a traditional office visit for certain non-emergency conditions. This is a fast, affordable, and secure way to receive an online diagnosis and treatment plan for common medical conditions"

Sources Bavafa, H., Hitt, L. M., Terwiesch, C. (2018). The impact of e-visits on visit frequencies and patient Health: Evidence from primary care. *Management Science. 64*(12):5461–80. https://doi.org/10.1287/mnsc.2017.2900; UMPC. (2023). E-Visits. https://www.upmc.com/campaigns/southcentral-pa/patient-portal/e-visits

F

False positive "A test result that indicates that a person has a specific disease or condition when the person actually does not have the disease or condition"

Source National Cancer Institute. (n.d.). false-positive test result. https://www.cancer.gov/publications/dictionaries/cancer-terms/def/false-positive-test-result

Functional ability "consists of the intrinsic capacity of the individual, relevant environmental characteristics and the interaction between them" and "is about having the capabilities that enable all people to be and do what they have reason to value"

Source World Health Organization. (2020, October 26). Healthy ageing and functional ability. https://www.who.int/news-room/questions-and-answers/item/healthy-ageing-and-functional-ability#:~:text=Functional%20ability%20consists%20of%20the,and%20the%20interaction%20between%20them.

Futures cone "illustrates the variety of possible futures that emanate from the current moment in time. Applied to strategy, analysis of the range of futures for any given issue can inform strategic planning"

Source Goal Atlas (n.d.). Strategy models & frameworks. https://goalatlas.com/futures-cone/

Fuzzy boundaries "Incorporates the uncertainty or the probability associated with "set membership." For example, "Are you in the forest or not?" This is not a boolean condition, but a highly complex boundary that changes with scale and time"; "arise from the way space is structured internally and how this relates to the external structure of space. Such boundaries can be effective in supporting functional differentiation of areas or the growth of areal identities and characters, but do not depend on the area being either spatially self contained or geometrically differentiated, or having clear spatial limits"

Sources Harbor, D., Geol 260—GIS & Remote Sensing. https://geol260.academic.wlu.edu/course-notes/fuzzy-logic-fuzzy-sets-conditional-inclusion-and-bayes-theorem/a-fuzzy-boundary/#:~:text=Fuzzy%20Boundaries,changes%20with%20scale%20and%20time.; Yang, T. (2007, January). The fuzzy boundary: the spatial definition of urban areas. 6th International Space Syntax Symposium. https://www.researchgate.net/publication/316236636_The_fuzzy_boundary_the_spatial_definition_of_urban_areas

G

Generative artificial intelligence "describes algorithms (such as ChatGPT) that can be used to create new content, including audio, code, images, text, simulations, and videos. Recent breakthroughs in the field have the potential to drastically change the way we approach content creation"; "is a type of artificial intelligence (AI) system capable of generating text, images, or other media in response to prompts. Generative AI models learn the patterns and structure of their input training data, and then generate new data that has similar characteristics"

Sources McKinsey & Company. (2023, January). What is generative AI?. https://www.mckinsey.com/featured-insights/mckinsey-explainers/what-is-generativ e-ai; Wikipedia. (2023, June 20). https://en.wikipedia.org/wiki/Generative_artificial_int elligence

Gerontechnology "is defined as an interdisciplinary field linking existing and developing technologies to the aspirations and needs of aging and aged adults. Gerontechnology supports "successful aging", and is a response to the combination of the aging of society and rapidly emerging new technologies"

Source Chen, L. K. (2020). Gerontechnology and artificial intelligence: Better care for older people. *Archives of Gerontology and Geriatrics*. 91, 104,252. https://doi.org/10.1016/j.archger.2020.104252. PMID: 32,979,638; PMCID: PMC7482590.

Geroscience "seeking to understand the genetic, molecular, and cellular mechanisms that make aging a major risk factor and driver of common chronic conditions and diseases of older people"

Source National Institute on Aging (n.d.). *Geroscience: The intersection of basic aging biology, chronic disease, and health*. https://www.nia.nih.gov/research/dab/geroscience-intersection-basic-aging-biology-chronic-disease-and-health

Geroscience approach "The Geroscience hypothesis posits that since aging physiology plays a major role in many—if not all—chronic diseases, therapeutically addressing aging physiology directly will prevent the onset or mitigate the severity of multiple chronic diseases"

Source American Federation for Aging Research (2023). What is geroscience?. https://www.afar.org/contact.

H

Health literacy "Personal health literacy is the degree to which individuals have the ability to find, understand, and use information and services to inform health-related decisions and actions for themselves and others. Organizational health literacy is the degree to which organizations equitably enable individuals to find, understand, and use information and services to inform health-related decisions and actions for themselves and others."

Source Centers for Disease Control and Prevention (2023). What is health literacy? https://www.cdc.gov/healthliteracy/learn/index.html

High quality data capture "Data quality refers to the state of qualitative or quantitative pieces of information...data is generally considered high quality if it is 'fit for [its] intended uses in operations, decision making and planning' ... data is deemed of high quality if it correctly represents the real-world construct to which it refers"

Source Wikipedia (2022). Data quality. https://en.wikipedia.org/wiki/Data_quality

High tech "High-tech, as opposed to high-touch, is when customers don't need human interaction to perform activities such as onboarding, ordering, and account management.https://en.wikipedia.org/wiki/High-touch High-tech services, such at chatbots, allow customers to get what they want on-demand"

Source Wikipedia. (2022). https://en.wikipedia.org/wiki/High-touch

Home care technologies "From emergency notification systems to videoconferencing platforms, home care technology helps clients retain their independence without sacrificing their safety"; "will enable individuals with chronic conditions to be more independent in their home, remain in their homes longer and be more engaged in the self-management of their conditions. Other areas impacted by technologies include communication, medication management, daily activities, cognitive fitness and social networking"

Sources Aetonix (n.d.). Home care technology: A new frontier in health care. https://aetonix.com/monitoring-home-patients/home-care-technology-new-frontier-health-care/; Hospital News (n.d.) Technology-enabled home care. https://hospitalnews.com/technology-enabled-home-care/

Homophily "the tendency for individuals who are socially connected in some way to display certain affinities, such as similarities in demographic background, attitudes, values, and so on"

Source American Psychological Association (2023). APA Dictionary of Psychology. https://dictionary.apa.org/homophily

Hypoaesthesia "An abnormality of the primary sensation that is mediated by peripheral nerves (pain, temperature, touch, vibration, joint position). The word hypoesthesia (or hypesthesia) refers to a reduction in cutaneous sensation to a specific type of testing"

Source National Library of Medicine. (n.d.). https://www.ncbi.nlm.nih.gov/medgen/6974#:~:text=Definition,from%20HPO%5D

I

Independence "The state or quality of being independent; freedom from the influence, control, or determination of another or others … The state or quality of being independent."

Source Collins (Dictionary) (2024). Independence. https://www.collinsdictionary.com/dictionary/english/independence

Informatics "studies the structure, behaviour, and interactions of natural and artificial systems that store, process, and communicate information"

Source Cambridge Dictionary (2023). https://dictionary.cambridge.org/dictionary/english/informatics

Informed consent "Consent is informed if, before giving it: the person received the information about the treatment that a reasonable person in the same circumstances would require to make a decision; and the person received responses to his/her requests for additional information about the treatment. The information must include the: nature of the treatment; expected benefits of the treatment; material risks and side effects of the treatment; alternative courses of action; and likely consequences of not having the treatment. Consent can be withdrawn at any time. Informed consent does not always need to be written, but can be oral or implied."

Source College of Nurses of Ontario, Practice Guidelines, Consent. https://www.cno.org/globalassets/docs/policy/41020_consent.pdf

Innovation-system fit "An innovation that fits with the organization's existing values, norms, strategies, goals, skill mix, supporting technologies, and ways of working is more likely to be assimilated"

Source Greenhalgh, T., Robert, G., Macfarlane, F., Bate, P., Kyriakidou, O. (2004). Diffusion of innovations in service organizations: Systematic review and recommendations. *Milbank Quarterly. 82*(4), 581–29. https://doi.org/10.1111/j.0887-378X.2004.00325.x

Integrated health "Brings conventional and complementary approaches together in a coordinated way. ... emphasizes multimodal interventions, which are two or more interventions such as conventional health care approaches (like medication, physical rehabilitation, psychotherapy), and complementary health approaches (like acupuncture, yoga, and probiotics) in various combinations, with an emphasis on treating the whole person rather than, for example, one organ system. Integrative health aims for well-coordinated care among different providers and institutions by bringing conventional and complementary approaches together to care for the whole person." See also: Whole person health: "Whole person health refers to helping individuals, families, communities, and populations improve and restore their health in multiple interconnected domains—biological, behavioral, social, environmental—rather than just treating disease. Research on whole person health includes expanding the understanding of the connections between these various aspects of health, including connections between organs and body systems."

Source National Center for Complementary and Integrative Health (2024). Complementary, Alternative, or Integrative Health: What's In a Name? https://www.nccih.nih.gov/health/complementary-alternative-or-integrative-health-whats-in-a-name

Integrated health services "integrated health service delivery": "The management and delivery of health services so that clients receive a continuum of preventive and curative services, according to their needs over time and across different levels of the health system"; "**Integrated services** are health services that are managed and delivered so that people receive a continuum of health promotion, disease prevention, diagnosis, treatment, disease-management, rehabilitation and palliative care services, coordinated across the different levels and sites of care within and beyond the health sector, and according to their needs throughout the life course."

Source World Health Organization, (2008), Technical Brief No.1, 2008, Integrated Health Services—What And Why?. https://terrance.who.int/mediacentre/data/sage/SAGE_Docs_Ppt_Apr2014/10_session_child_health_services/Apr2014_session10_integrated_health_services.pdf; World Health Organization, Services Organization and Integration. https://www.who.int/teams/integrated-health-services/clinical-services-and-systems/service-organizations-and-integration

Interdisciplinary "involving two or more different subjects or areas of knowledge"

Source Cambridge Dictionary (2023). https://dictionary.cambridge.org/dictionary/english/informatics

Interoperability "is the ability of different systems, devices, applications or products to connect and communicate in a coordinated way, without effort from the end user. Functions of interoperable components include data access"

Source Lewis, S. (2023). TechTarget. https://www.techtarget.com/searchapparchitecture/definition/interoperability

Intersectionality "the ways in which systems of inequality based on gender, race, ethnicity, sexual orientation, gender identity, disability, class and other forms of discrimination "intersect" to create unique dynamics and effects"; "a framework that describes how our overlapping social identities relate to social structures of racism and oppression. Intersectionality merges many identity markers, including race, class, gender, sexual orientation, age, ethnicity, religion, disability, and more, to create a more truthful and complex identity"

Sources Center for Intersectional Justice (n.d.). https://www.intersectionaljustice.org/what-is-intersectionality; University of British Columbia (2021). Intersectionality: What is it and why it matters. https://vpfo.ubc.ca/2021/03/intersectionality-what-is-it-and-why-it-matters/

Intersectional theory "Intersectional theory views the categories of intersecting relations such as race, gender, social class, sexuality, ability, and age as interrelated and mutually shaping one another. Through taking these intersecting factors into consideration, it paves the way of understanding and explaining complexity in individuals, the world, and in human experience"

Source Guy-Evans, O. SimplySociology (2023). https://simplysociology.com/intersectional-theory.html

Interstitial space "Interstitial spaces are informal settings where interactions among actors from different organizational fields are more likely to succeed despite the institutional complexity that they face"

Source Villani, E., & Phillips, N. (2021). Formal organizations and interstitial spaces: Catalysts, complexity, and the initiation of cross-field collaboration. *Strategic Organization*, *19*(1), 5–36. https://doi.org/10.1177/1476127019897235

Intrinsic capacity "Intrinsic capacity, a crucial concept in healthy ageing, is defined by WHO as 'the composite of all the physical and mental capacities that an individual can draw on at any point in time'. Vitality capacity is considered the underlying physiological determinant of intrinsic capacity"

Source Bautmans, I., Veerle Knoop, V., Thiyagarajan, J. A., Maier, A. B., Beard, J. R., Freiberger, E., Belsky, D., Aubertin-Leheudre, M., Mikton, C., Cesari, M., https://www.thelancet.com/journals/lanhl/article/PIIS2666-7568(22)00200-8/fulltextSumi, Y., Diaz, T., & Banerjee, A. (2022). WHO working definition of vitality capacity for healthy longevity monitoring. *The Lancet*. https://doi.org/10.1016/S2666-7568(22)00200-8

J

Journey mapping "a UX visualization document that showcases the steps that a user takes in a process to accomplish a goal. Personas are created with information gathered from user and stakeholder interviews. As a result of these activities, you can identify the most important functionality an audience needs"; "are used to map the relationship between a customer and an organization over time and across all channels on which they interact with the business. Design teams use customer journey maps to see how customer experiences meet customers' expectations and find areas where they need to improve designs"

Sources Yale University (2023). User journey maps. https://usability.yale.edu/understanding-your-user/user-journey-maps; Interaction Design Foundation. (n.d.), Customer journey maps. https://www.interaction-design.org/literature/topics/customer-journey-map

K

Knowledge mobilization "is an umbrella term encompassing a wide range of activities relating to the production and use of research results, including knowledge synthesis, dissemination, transfer, exchange, and co-creation or co-production by researchers and knowledge users"

Source Social Sciences and Humanities Research Council (2023). Guidelines for effective knowledge mobilization. https://www.sshrc-crsh.gc.ca/funding-financement/policies-politiques/knowledge_mobilisation-mobilisation_des_connaissances-eng.aspx#:~:text=and%20related%20activities.-,What%20is%20knowledge%20mobilization%3F,by%20researchers%20and%20knowledge%20users.

Knowledge translation "is defined as a dynamic and iterative process that includes synthesis, dissemination, exchange and ethically-sound application of knowledge to improve the health of Canadians, provide more effective health services and products and strengthen the health care system. This process takes place within a complex system of interactions between researchers and knowledge users which may vary in intensity, complexity and level of engagement depending on the nature of the research and the findings as well as the needs of the particular knowledge user"

Source Canadian Institutes of Health Research (2016). About us. https://cihr-irsc.gc.ca/e/29418.html#2

L

Learning health system (LHS) is one where "science, informatics, incentives, and culture are aligned for continuous improvement and innovation, with best practices seamlessly embedded in the care process, patients and families active participants in all elements, and new knowledge captured as an integral by-product of the care experience"

As define by the Institute of Medicine Committee on the Learning Health Care System in America; Institute of Medicine. Best Care at Lower Cost: The Path to Continuously Learning Health Care in America. Smith M, Saunders R, Stuckhardt L, McGinnis JM, editors. Washington (DC): National Academies Press (US); 2013 May 10. PMID: 24,901,184.

Life-cycle health technology assessment "is the standardization of data collected and methods needed to inform life-cycle appraisal, re-appraisal, and de-adoption of health technologies. The framework includes: managed access for conditional support subject to collecting real world evidence; collecting real world data and concatenating that data with external data; generating real world evidence for decisions, and interpreting the evidence to decide on adoption, disinvesting from managed access, or continuing to collect data"

Source Regier, D. A., Pollard, S., McPhail, M., Bubela, T., Hanna, T. P., Ho, C., Lim, H. J., Chan, K., Peacock, S. J., Weymann, D. A. (2022). Perspective on life-cycle health technology assessment and real-world evidence for precision oncology in Canada. *npj Precision Oncology*, *6*(76). https://doi.org/10.1038/s41698-022-00316-1

Lived experience "Personal knowledge about the world gained through direct, first-hand involvement in everyday events rather than through representations constructed by other people. It may also refer to knowledge of people gained from direct face-to-face interaction rather than through a technological medium"

Source Oxford Reference (2023). https://www.oxfordreference.com/display/10.1093/oi/authority.20110803100109997;jsessionid=681205F00BB9C5ADFD01AA54FFF330F5

Locomotor capacity "is broadly described as a person's bodily capacity or physical capacity to move from one place to another … Four main characteristics proposed in a recent publication include a function of joints, bones, reflexes and muscle strength"

Source Veronese, N., Honvo, G., Amuthavalli Thiyagarajan, J., Rizzoli, R., Cooper, C., Bruyère, O., Mikton, C., Sumi, Y., Diaz, T., Reginster, J-Y., & Banerjee, A. (2022). Attributes and definitions of locomotor capacity in older people: a World Health Organisation (WHO) locomotor capacity working group meeting report. *Aging Clinical and Experimental Research, 34*, 481–3 https://doi.org/10.1007/s40520-022-02080-5

M

Marginalisation "Also referred to as social exclusion, occurs when certain groups of people get denied access to areas of society. Many factors can lead to this denial of access to institutions and opportunities, including historical bias and lack of funding. Marginalized people don't necessarily belong to one particular demographic: Marginalization occurs due to ethnicity, gender, sexual orientation, disability status, socioeconomic level, and age. Marginalized groups are often at a disadvantage when it comes to obtaining health care, decent education, and employment that would improve their well-being. Marginalization puts people in powerless positions based on one or several dimensions of exclusion"… Economic, political, and social marginalization.

Source MasterClass (2022). What Is marginalization? Types, causes, and effects. https://www.masterclass.com/articles/marginalization-explained

Meta-analysis "a quantitative technique for synthesizing the results of multiple studies of a phenomenon into a single result by combining the effect size estimates from each study into a single estimate of the combined effect size or into a distribution of effect sizes"

APA Dictionary of Psychology. (2023). https://dictionary.apa.org/meta-analysis

Micro levels of organizations "involves direct intervention and one-to-one support of individuals, families and small groups navigating personal and social challenges"

Source St John, M., and Freeman, V. (2022). Macro, micro and mezzo social work: What's the difference? *Forbes Advisor*. https://www.forbes.com/advisor/education/macro-micro-mezzo-social-work/

Mezzo/meso levels of organizations "promotes social development and cohesion within larger groups, organizations and communities"; "focus on how individuals function within small systems such as neighborhoods, schools, or other organizations" and "Mezzo social work is working with people in the systems that they live, learn, work, and socialize in. It is based on an understanding that individuals act and respond differently in different situations and with different people"

Sources St John, M., and Freeman, V. (2022). Macro, micro and mezzo social work: What's the difference? *Forbes Advisor*. https://www.forbes.com/advisor/education/macro-micro-mezzo-social-work/; DegreeChoices.com (2023). Mezzo social work: A comprehensive guide. https://www.degreechoices.com/mezzo-social-work/

mHealth (mobile health) technologies "is a subset of electronic health and refers to the use of portable wireless devices capable of transmitting, storing, processing, and retrieving real-time and non-real-time data between patients and healthcare providers" and "increasing efficiency, reducing costs, and providing healthcare in the least time with the minimum risk are the most important benefits of mHealth … include multiple functionalities such as education for patients and healthcare providers, diagnostic support, operative planning, postoperative care, and follow-up management"

Source Zakerabasali, S., Ayyoubzadeh, S. M., Baniasadi, T., Yazdani, A., & Abhari, S. (2021). Mobile health technology and healthcare providers: Systemic barriers to Adoption. *Healthcare Informatics Research, 27*(4), 267–78. https://doi.org/10.4258/hir.2021.27.4.267

Morbidity "refers to having a disease or a symptom of disease, or to the amount of disease within a population. Morbidity also refers to medical problems caused by a treatment"

Source National Cancer Institute (n.d.). https://www.cancer.gov/publications/dictionaries/cancer-terms/def/morbidity

Multidisciplinary rehabilitation "Multidisciplinary is when many different disciplines work together toward a common goal"; "Most rehabilitation teams hold weekly, biweekly, or monthly meetings, depending on the setting"

Source John Hopkins Medicine (2023). Overview of the PM&R Treatment Team. https://www.hopkinsmedicine.org/health/treatment-tests-and-therapies/overview-of-the-pmr-treatment-team

N

Nursing home "Nursing homes, also called skilled nursing facilities, provide a wide range of health and personal care services. Their services focus on medical care more than most assisted living facilities. These services typically include nursing care, 24-h supervision, three meals a day, and assistance with everyday activities. Rehabilitation services, such as physical, occupational, and speech therapy, are also available"

Source National Institute on Aging (n.d.). Residential facilities, assisted living, and nursing homes. https://www.nia.nih.gov/health/residential-facilities-assisted-living-and-nursing-homes

P

Path dependency "Path dependence is the notion that decisions people are faced with now depend on previous decisions or choices made and past knowledge trajectory, so they are limited by the current competence base. In other words, our current decision-making situations are determined by history—what happened in the past has a huge influence on today's strategic planning"

Source Market Business News (2023). What is path dependence? Definition and meaning. https://marketbusinessnews.com/financial-glossary/path-dependence-definition-meaning/

Persona "A persona (also user persona, customer persona, buyer persona) in user-centered design and marketing is a fictional character created to represent a user type that might use a site, brand, or product in a similar way"

Source Wikipedia (2023). Persona. https://en.wikipedia.org/wiki/Persona_(user_experience)

Physiatry "is the medical specialty that emphasizes the prevention, diagnosis, treatment and rehabilitation of people disabled by disease, disorder or injury"

Source Association of American Physiatrists. (n.d.). What is physiatry?. https://www.physiatry.org/page/WhatIsPhysiatry#:~:text=Physiatry%20is%20the%20medical%20specialty,by%20disease%2C%20disorder%20or%20injury.

Physiological age "a measurement of the level of development, health, or deterioration of an individual in terms of functional norms for various body systems"

Source American Psychology Association (2023). APA Dictionary of Psychology. https://dictionary.apa.org/physiological-age

Policy landscapes "A landscape is all the features that are important in a particular situation"

Source Collins Dictionary (2023). https://www.collinsdictionary.com/dictionary/english/political-landscape

Primary care "Primary health care serves a dual function in the health care system: a direct provision of first-contact services (by providers such as family physicians, nurse practitioners, pharmacists, and telephone advice lines); and a coordination function to ensure continuity and ease of movement across the system, so that care remains integrated when patients require more specialized services (with specialists or in hospitals, for example)"

Source Government of Canada (2012). About primary health care. https://www.canada.ca/en/health-canada/services/primary-health-care/about-primary-health-care.html#a1

Psychological capacity "the maximum ability of an individual to receive or retain information and knowledge or to function in mental or physical tasks" and "the potential of an individual for intellectual or creative development or accomplishment"

Source American Psychological Association (2023). *APA Dictionary of Psychology*. https://dictionary.apa.org/capacity

Purkinje neurons "are a unique type of neuron-specific to the cerebellar cortex. They are remarkable (and instantly recognizable) for their massive, intricately branched, flat dendritic trees, giving them the ability to integrate large amounts of information and learn by remodeling their dendrites. As an important part of the cerebellar circuits, Purkinje cells are necessary for well-coordinated movement and other areas of function such as cognition and emotion"

Source National Library of Medicine. (n.d.). https://www.ncbi.nlm.nih.gov/books/NBK545154/#:~:text=Purkinje%20cells%20are%20a%20unique,learn%20by%20remodeling%20their%20dendrites.v

R

Resource system "Resource-System scope is a key functionality for capacity planning in manufacturing. For a given system, it defines the resource scope on which the system (and operations of the system) can be executed. This scope is a prerequisite for balancing operations on resources"

Source Catia help. (n.d.). About resource-system scope. https://catiahelp.azurewebsites.net/English/MlbUserMap/mlb-c-Scope.htm

Royal Commissions "are a form of official inquiry into matters of public concern" and "are necessary from time to time to flush out misfeasance and malfeasance and to examine matters of major public concern. Some royal commissions have produced significant reports"

Source Fox, P. (2023). Royal Commissions. *The Canadian Encyclopedia.* https://www.thecanadianencyclopedia.ca/en/article/royal-commissions

S

Scoping review "A scoping review of a body of literature can be of particular use when the topic has not yet been extensively reviewed or is of a complex or heterogeneous nature … They are commonly undertaken to examine the extent, range, and nature of research activity in a topic area; determine the value and potential scope and cost of undertaking a full systematic review; summarize and disseminate research findings; and identify research gaps in the existing literature … As it provides a rigorous and transparent method for mapping areas of research, a scoping review can be used as a standalone project or as a preliminary step to a systematic review"

Source Pham, M. T., Rajić, A., Greig, J. D., Sargeant, J. M., Papadopoulos, A., McEwen, S. A. (2014). A scoping review of scoping reviews: Advancing the approach and enhancing the consistency. *Research Synthesis Methods*, *5*(4), 371–5. https://doi.org/10.1002/jrsm.1123

Seniors "the population of seniors, i.e. the population of individuals aged 65 and over"

Source Turcotte, M., & Schellenberg, G. (2007). A portrait of seniors in Canada: Introduction. https://www150.statcan.gc.ca/n1/pub/89-519-x/89-519-x2006001-eng.htm

Siloed "Organizational silos describe the isolation that occurs when employees or entire departments within an organization do not want to, or do not have the adequate means to share information or knowledge with each other"

Source Ismail, K. (2018). 5 signs your organization is siloed. *CMSWire*. https://www.cmswire.com/leadership/5-signs-your-organization-is-too-siloed/

Slack resources "Slack is a pool of organizational resources in excess of the minimum necessary to produce a given level of organizational output … can provide a cushion that allows organizations to adjust successfully to internal pressures as well as to initiate strategy with respect to the external environment"

Source Zinn, J., & Flood, A. B. (2009). Commentary: Slack resources in health care organizations—Fat to be trimmed or muscle to be exercised? Health Services Research, *44*(3), 812–20. https://doi.org/10.1111/j.1475-6773.2009.00970.x

Social determinants "are the non-medical factors that influence health outcomes. They are the conditions in which people are born, grow, work, live, and age, and the wider set of forces and systems shaping the conditions of daily life. These forces and systems include economic policies and systems, development agendas, social norms, social policies and political systems"; "have an important influence on health inequities"

*Source*WHO (2023). Social determinants of health. https://www.who.int/health-topics/social-determinants-of-health#tab=tab_1

Soft periphery "the organizational structures and systems required for the full implementation of the innovation"

Source Greenhalgh, T., Robert, G., Macfarlane, F., Bate, P., & Kyriakidou, O. (2004). Diffusion of innovations in service organizations: Systematic review and recommendations. *Milbank Quarterly*, *82*(4), 581–29. https://doi.org/10.1111/j.0887-378X.2004.00325.x

Stepped-wedge cluster "involves the sequential transition of clusters (such as hospitals, public health units or communities) from control to intervention conditions in a randomized order"; "a logistic constraint typically prevents the simultaneous treatment of some participants, and instead, all or most participants receive the treatment in waves or 'steps'"

Sources Hemming, K., & Taljaard, M. (2020, June). Reflection on modern methods: when is a stepped-wedge cluster randomized trial a good study design choice? International Journal of Epidemiology. *49*(3), 1043–52. https://doi.org/10.1093/ije/dyaa077; Wikipedia (2023). Stepped-wedge trial. https://en.wikipedia.org/wiki/Stepped-wedge_trial

Structural determinants "include the governing processes, economic and social policies that can affect everything from sick leave policy to housing or education which are the intermediary step to unequal distribution of power, prestige and resources"

*Source*May, V. M. (2021). Structural vs social determinants of health and social risk vs social needs of health: Precision matters for accountability for effective change. HHS.Gov. https://ncvhs.hhs.gov/wp-content/uploads/2021/08/Presentation-Panel-4-UCLA-Vickie-Mays-508.pdf

Synergy "the combined power of a group of things when they are working together that is greater than the total power achieved by each working separately"

Source Cambridge Dictionary (2023). https://dictionary.cambridge.org/dictionary/english/synergy

Systematic review "typically involve a detailed and comprehensive plan and search strategy derived a priori, with the goal of reducing bias by identifying, appraising, and synthesizing all relevant studies on a particular topic. Often, systematic reviews include a meta-analysis component which involves using statistical techniques to synthesize the data from several studies into a single quantitative estimate or summary effect size"

Source Uman, L. S. (2011). Systematic reviews and meta-analyses. *Canadian Academy of Child and Adolescent Psychiatry. 20*(1), 57–59. PMID: 21,286,370; PMCID: PMC3024725.

T

Tension for change "is an important concept for business leaders looking to improve performance in their companies. It is a mechanism to create the energy and motivation needed to mobilize human beings into action"

Source O'Connor, T. (2019). Creating tension for change. Results Business Consultants. https://blog.unleashresults.com/creating-tension-for-change#:~:text=Tension%20for%20change%20is%20an,is%20the%20work%20of%20leaders.

Transdisciplinary "is generally defined by the inclusion of non-academic stakeholders in the process of knowledge production"

Source Rigolot, C. (2020). Transdisciplinarity as a discipline and a *way of being*: Complementarities and creative tensions. *Humanities & Social Sciences Communications*, 7, 100. https://doi.org/10.1057/s41599-020-00598-5

Trialability "The ability to test the intervention on a small scale in the organization, and to be able to reverse course (undo implementation) if warranted"

Source University of Colorado (2023). Dissemination & implementation models in health. https://dissemination-implementation.org/tool/

U

Umbrella review "are systematic collections and assessments of multiple SRMAs [systematic reviews and meta-analyses] done on a specific research topic"; "is a cluster of existing systematic reviews on a shared topic … are also known as an overview of reviews. [and] … are 'overarching reviews' that 'agreggrat[e] findings from several reviews that address specific questions' … 'each umbrella review focuses on a broad condition or problem for which there are two or more potential interventions and highlights reviews that address these potential interventions and their results"

Sources Belbasis, L., Bellou, V., & Ioannidis, J. P. A. (2022). Conducting umbrella reviews. *bmj medicine.* https://bmjmedicine.bmj.com/content/1/1/e000071; Gerstein Science Information Centre (2023). Knowledge syntheses: Systematic & scoping reviews, and other review types. University of Toronto. https://guides.library.utoronto.ca/c.php?g=713309&p=5083951

Usability studies "Usability … extent to which a product can be used by specified users to achieve specified goals with effectiveness, efficiency and satisfaction in a specified context of use"; "Usability testing refers to evaluating a product or service by testing it with representative users. Typically, during a test, participants will try to complete typical tasks while observers watch, listen and takes notes. The goal is to identify any usability problems, collect qualitative and quantitative data and determine the participant's satisfaction with the product"

Sources Edith Cowan University. (2023). Research Methodologies for the Creative Arts & Humanities: Usability studies. https://ecu.au.libguides.com/research-methodologies-creative-arts-humanities/usability-studies#:~:text=Definitions%3A,2006); Usability.Gov. (2023). Usability testing. https://www.usability.gov/how-to-and-tools/methods/usability-testing.html

V

Value of information "is a decision analytic method for quantifying the benefit of acquiring additional information to support such analyses that can be used to help in a wide range of manufacturing decisions"; "is a method for estimating the expected monetary gain from reducing uncertainty in the input parameters of a decision analytic model"

Sources Zabeo, A., Keisler, J. M., Hristozov, D., Marcomini, A., Linkov, I. (2019). Value of information analysis for assessing risks and benefits of nanotechnology innovation. *Environmental Sciences Europe*, *31*(11). https://doi.org/10.1186/s12302-019-0194-0; Fernando Alarid-Escudero, F., Knowlton, G., Enns, E., & the DARTH Team. (2021, June 30). Value of information analysis. The R project for statistical computing. https://cran.r-project.org/web/packages/dampack/vignettes/voi.html

Virtual care "virtual care as any kind of interaction between patients and members of their healthcare team that occurs remotely … might include email, secure text messaging, phone calls, remote monitoring or face-to-face video appointments—basically any form of care where the patient and the provider are not physically in the same room"

Source Women's College Hospital (2019). Virtual care: Technology-enabled care connects patients and their healthcare providers. http://wearewomensreport2019.womenscollegehospital.ca/virtual-care/

W

Web portals "A web portal is a specially designed website that often serves as the single point of access for information. It can also be considered a library of personalized and categorized content. A web portal helps in search navigation, personalization, notification and information integration, and often provides features like task management, collaboration, and business intelligence and application integration"

Source Rouse, M. (2017). What does web portal mean? Technopedia. https://www.techopedia.com/definition/17352/web-portal

Numbers

10,000-m view "a description or view of a situation or problem that provides very general information, but no details"

Source Cambridge Dictionary (2023). https://dictionary.cambridge.org/dictionary/english/10000-foot-view

Index

A

Activities of Daily Living (ADL), 62, 74, 93, 96, 97, 105
Acute care, 43, 61, 63, 96, 106, 174
Ada, 129
Adoption, 6, 7, 46, 65–67, 78, 127, 139, 142, 144, 160, 165, 172–175, 177, 187, 189, 213, 215
Adverse health, 41
Adverse medical events, 13, 43, 44
Advisory panel on healthcare innovation, 2015, 172
Advocacy, 32, 106, 156, 159, 160, 164
Age, 3, 5, 11, 15, 17, 18, 21, 26, 29, 41–43, 52, 53, 58, 62, 93, 97, 101, 106, 107, 110, 127, 142, 152, 156, 157, 163, 172, 177, 184, 193, 202, 203, 211, 214, 216, 218
Ageing populations, 15
Age in place, 64, 182
Ageism, 26, 156, 163, 188, 194
AgeTech, 3–8, 18–22, 28–32, 35–37, 43, 44, 57, 64, 94, 95, 97, 98, 100, 102, 106–109, 125, 131, 152–155, 158–168, 171, 172, 174–178, 181–187, 189, 193–199, 201
AGE-WELL, 71
AGE-WELL network, 71
Aging, 5–8, 11, 13–15, 17–21, 28–30, 33, 43, 44, 47, 49, 57, 59, 61, 63, 64, 91, 93, 94, 97, 105, 152, 153, 161, 164, 176, 181–183, 185–189, 194, 195, 198, 201, 202, 208, 214, 215
Alberta, 98, 108, 129, 203, 204
Alzheimer's disease, 58
A new perspective on the health of Canadians, 29
App, 50, 70, 94, 103, 109, 128, 129, 143, 158, 168, 177, 187, 197, 198
APPTA, 21
Arthritis, 41, 58, 73, 77, 139, 203
Artificial Intelligence (AI), 49, 52, 95, 98, 129, 159, 201, 202, 207, 208
Assessment Urgency Algorithm (AUA), 143, 146
Assistive devices, 11, 12, 44, 50–52, 101, 110, 196
Assistive technologies, 42, 44, 54, 70, 71, 94, 98, 100, 110, 185
At-risk orphan subpopulations, 152
At-risk populations, 49
Augmentative and Alternative Communication (AAC), 70
Australia, 27, 59, 62, 94, 101, 205
Australian policy handbook, 32
Autonomy, 22, 29, 43, 160, 161, 164, 166, 171, 175, 177, 178, 196, 202

B

Baby boomer, 95
Babylon app, 129
Berlin, 129
Biological age, 17, 18, 202
Biological mechanisms of aging, 17
Boomer, 156

D. Juzwishin et al., *AgeTech Innovations in Healthcare for Older Adults*, Synthesis Lectures on Technology and Health, https://doi.org/10.1007/978-3-031-86885-6

Brazil, 125
Butterfly Household Model of Care, 60
Butterfly Model, 60

C
Canada, 5, 17, 19, 21, 27, 34, 51, 54, 59, 61–65, 94, 96, 98, 100–102, 105–107, 110, 125, 127, 141, 145, 160, 161, 163, 171, 172, 176, 177, 184, 204, 213, 216, 217
Canada Health Act, 34
Canada's Agency for Drugs in Technology and Health (CADTH), 99, 106, 109, 110, 176
Canadian Institute for Health Information, 61
Canadian Institutes of Health Research (CIHR), 17, 213
Care, 3, 4, 7, 15, 17, 19–22, 26, 27, 30, 37, 41–46, 49–54, 57–60, 62–70, 76, 78, 91, 93, 94, 96–98, 105–109, 119–131, 139–145, 151, 153, 158–162, 164, 165, 168, 174, 175, 177, 178, 181, 182, 184–186, 189, 198, 199, 202, 204, 208, 210, 211, 213, 215, 216, 220, 221
Caredove, 141
Caregivers, 5, 18, 20, 21, 42–49, 52–54, 59, 65, 77, 97, 102, 106, 107, 110, 119, 128, 130, 142, 146, 173, 177, 187
Care recipients, 43, 45–53
Centre for Digital Health Evaluation (CDHE), 176
Chief Public Health Officer, 19
Chronic conditions, 17, 18, 22, 58, 97, 105, 109, 121, 128, 193, 208, 209
Chronological age, 17, 202, 203
CIHR Institute of Aging Strategic Plan 2023–2028, 17, 18
Clinical Assessment Protocols (CAPs), 68
Cochrane Database of Systematic Reviews, 110
Cochrane Library, 110
Co-created, 161, 162, 167, 168, 171, 177
Co-creation, 74, 77, 78, 155, 161, 162, 178, 186, 212
Co-design, 47, 67, 128, 130, 186–188
Co-designed, 188
Cognition, 14, 15, 68, 97, 101, 141, 204, 206, 217
Cognitive abilities, 11, 44, 92
Cognitive health, viii
Commonwealth Fund, 27, 204
Communication, 13, 19, 21, 28, 45–52, 54, 65, 70, 71, 77, 91, 119, 121–123, 125, 129, 131, 141, 142, 144, 145, 164, 173, 174, 177, 185, 187, 202, 206, 209, 219
Co-morbidities, 15, 22, 163, 181
Computer Interactive Reminiscence and Conversation Aid system (CIRCA), 71
Computerized Provider Order Entry (CPOE), 141
Computers, 6, 30, 45–47, 49, 71, 75, 98, 101, 103, 126, 155, 157, 177, 201
Confidentiality, 22, 74, 76, 145, 160, 162, 177, 186
COVID-19, 3, 11, 28, 37, 42, 49, 57, 60, 61, 63, 64, 66, 70, 73, 77, 105, 109, 126, 145, 156–159, 163, 167, 176, 184, 185, 198
COVID-19 pandemic, 3, 42, 57, 61, 64, 66, 70, 73, 77, 105, 109
Culture, 51, 78, 154, 173, 186, 188, 213

D
Data, 29, 33, 36, 45, 49–52, 54, 61, 67, 70, 73, 76, 100, 101, 141–146, 158–162, 168, 171, 172, 178, 183, 186, 195, 198, 202, 205, 208, 209, 211, 213, 215, 219, 220
David Naylor, 109, 172
David Sheard, 60
Death, 18, 19, 60, 61, 161, 163, 199, 204
Decade of Healthy Ageing (2021–2030), 3, 53, 185, 194
Delivery systems, 4–6, 27, 28, 130, 154, 163, 168, 171, 196
Dementia, 11, 14, 41, 47, 54, 60, 68, 70, 71, 74–76, 102, 107, 109, 160, 188
Denmark, 51, 59, 61, 160
Design for Dignity (D4D) Model, 53
Determinants of health, 19, 157
Dexterity, 13, 47, 50, 99, 176
Diabetes, 13, 19, 41, 203
Digital devices, 14, 42, 46, 72, 165

Digital divide, 126, 151, 155, 167, 185, 186, 193
Digital gaming, 73
Digital Medicine Society (DiME), 43, 165
Digital technologies, 3, 5, 6, 12, 13, 15, 19, 21, 28, 29, 31, 32, 37, 43, 45, 52, 54, 155, 157, 159, 161, 163–165, 173, 176, 189, 201, 205
Digital transformation, 6, 20
Dignity, 27, 29, 43, 96, 153, 154, 157, 171, 175, 182–186, 189, 193, 195, 196
Disability, 17, 19, 29, 54, 93–95, 100, 110, 143, 156, 159, 160, 193, 201, 211, 214
Disease, 15, 17–19, 30, 43, 93, 94, 97, 100, 105, 106, 110, 153, 158, 159, 181, 193, 195, 202, 203, 206–208, 210, 215, 216
Diversity, 26, 50, 62, 96
Domains of human functioning, 44
Domestic violence, 25, 26, 163

E
Early-stage dementia, 11, 160
Economic security, 27
eFI, 143, 146
Elder abuse, 62, 163, 164, 197
Eldercare, 60
Elder orphans, 151
Electronic health, 125, 141, 144, 215
Electronic health information technology, 125
Electronic Health Records (EHRs), 7, 8, 65, 67, 125, 128, 139–146, 160, 171
Electronic health research, 125
Electronic Medical Record (EMR), 164
Engagement, 21, 65, 67, 71, 77, 78, 106, 126, 172, 173, 186–188, 213
English Poor Law of 1601, 60
Environment, 6, 17, 18, 29, 34, 35, 59, 63, 66, 95, 99, 101, 106, 154, 175, 186, 188, 201, 203–206, 218
EnvisAge, 183
Essential tremor, 50, 206
Ethnicity, 22, 62, 211, 214
Europe, 125, 220
E-visits, 125, 128–130, 206
Exoskeletons, 98–102, 107, 108

F
4Cs, 121
Finland, 157, 160
First Nations, 158
First Nations Health Authority (FNHA), 158
Frailty, 13, 14, 41, 43–45, 49, 51–54, 93, 125, 143, 145, 146, 156, 164, 177, 181, 195
Frailty Index (FI), 67, 143
Framing, 32, 33, 36, 196
France, 27, 157
Functional ability, 17, 18, 93, 94, 97, 100, 102, 207
Future, 3, 7, 8, 11, 14, 17, 36, 43, 46, 49, 52, 54, 70, 95, 108, 127, 128, 145, 165, 175, 178, 182, 184, 185, 189, 198, 206, 207
Futures cone, 36, 207
Futures literacy, 36

G
Gatekeeper, 100, 106, 122
Gender identity, 62, 211
Generative artificial intelligence, 12, 194, 196, 207
George E. P. Box, 31
Germany, 27, 59, 61, 94, 95
Gerontechnology, 5, 208
Gerontologist, 26
Geroscience, 17, 208
Glenrose Rehabilitation Hospital, 98, 106, 108, 109
Glenrose Rehabilitation Hospital Strategic Plan, 109
Global Burden of Disease, Injuries and Risk Factors Study, 94
Greece, 157
Green House Model, 60
Grey divide, 127

H
H3 [**h**ealth care and **h**ealth service delivery for the **h**ome environment], 44, 52, 53
Harold Lasswell, 25
Health, 3–8, 11, 12, 14, 15, 17–22, 25, 27–29, 34, 35, 37, 41, 43–45, 51–53, 57, 61–63, 65, 67, 68, 70, 72, 76–78, 91,

93–97, 99, 101, 103, 104, 106, 108–110, 119, 121–123, 125–131, 139–146, 151, 153, 154, 157–162, 165, 166, 168, 172–177, 181–183, 186, 187, 194–197, 201, 203, 204, 206, 208, 210, 211, 213–216, 218, 219
Health care, 3–5, 7, 20, 25, 27–31, 34, 35, 37, 41, 43, 67, 93, 96, 99, 102, 106, 109, 110, 119, 121, 128, 130, 141, 142, 145, 146, 151, 152, 155, 157–165, 167, 168, 173, 175, 177, 183–185, 188, 193, 195, 202–204, 209, 210, 213, 214, 216, 218
Health care delivery, 6, 8, 15, 28–30, 35, 130, 145, 154, 158, 159, 167, 168, 171, 176, 184, 194–196
Health care delivery system, 3–5, 7, 8, 25, 27, 29, 33, 152, 154, 155, 158, 159, 162, 164, 165, 167, 171, 173, 181, 183–186, 189, 195, 196
Healthcare interventions, 30, 31
Health care professionals, 19
Health care provider, 3, 17–20, 22, 28, 30, 34, 41, 44, 92, 95, 96, 98, 108, 122, 123, 128, 141, 146, 154, 155, 157–166, 175, 176, 178, 186
Health care reform, 7, 8, 21, 29, 34, 36, 175, 183
Health care setting, 8, 119, 123, 146, 160, 187
Health care standards, 20
Health care system, 5–8, 20, 21, 25, 27, 28, 30, 33, 60, 61, 68, 78, 95, 108, 110, 122–124, 126, 142, 153, 154, 156, 157, 160, 164, 167, 171–174, 181, 186, 193–196, 213, 216
Health care teams, 20, 120, 141
Health delivery organizations, 20
Health outcomes, 5, 29, 31, 68, 119, 121, 218
Health promotion, 5, 153, 159, 182, 193, 195, 197, 210
Health services delivery, viii
Health Standards Organization (HSO), 20
Health status, 29, 68, 73, 107, 146, 155, 157, 165, 204
Health system sustainability, 33
Health technology, 19, 33, 46, 54, 99, 106, 108, 110, 141, 173, 176, 213, 215
Health Technology Assessment (HTA), 7, 8, 28, 33, 165, 173, 176, 213
Healthy aging, 18, 76, 142, 175, 194
Healthy lifestyle, 21, 198
Hearing, 11, 13, 15, 44, 50, 91
High income countries, 64
High tech, 30, 209
High touch, 30
Home care, 5, 7, 11, 20, 27, 43–46, 50–54, 64, 92, 96, 119, 122, 123, 140, 146, 182, 185, 187, 209
Home environments, 14, 42, 43, 46, 48, 50, 53, 67, 125
Homeless, 151–153, 155, 158, 159, 161, 162, 166, 181
Homelessness, 153, 158, 159, 161, 162, 168
Humility, 77, 158, 163, 168, 186

I

I CAN Centre for Assistive Technology Services, 98
ICT tool—Systems for Person-centered Elder Care (SPEC), 69
Immigrants, 62, 188, 197
Impairments, 29, 44, 62, 67, 70, 71, 74, 77, 99, 104, 155, 158
Independence, 18, 27, 43, 67, 94, 96, 97, 174, 177, 178, 182, 209
Indigenous, 62, 127, 158, 197
Industrial robotics: Insights into the sector's future growth dynamics, 99
Information Technology (IT), 68
Innovation, 5, 7, 8, 28, 30–33, 35, 37, 46, 51, 54, 78, 108, 155, 158–160, 162, 166, 167, 172–178, 181, 183–187, 189, 195, 197, 204, 206, 210, 213, 218, 220
Innovation forum, 53, 54
Innovation forum on H3 technologies and systems, 53
Innovation pathway, 173
Innovators, 5, 8, 18, 20, 29, 33, 35–37, 43, 44, 46, 54, 78, 106, 108, 109, 155, 158, 159, 161, 163–166, 168, 174, 176–178, 183, 184, 186, 193
Interface, 45–48, 51, 106, 164
International Federation of Robotics (IFR), 98, 110

Internet, 6, 28, 49, 67, 71, 78, 153, 155, 157–159, 167, 168, 171, 186
InterRAI-LTCF, 68–70
InterRAI Minimum Data Set (MDS 2.0) assessment system, 68
Intersectionality theory, 153
Interstitial space, 151, 154, 181, 182, 211
Intrinsic capacity, 17, 18, 207, 212
Isolation, 15, 22, 28, 42, 151, 158, 218
Italy, 60, 103

J

Japan, 59
John W. Kingdon, 32

K

Knowledge, 7, 14, 18, 19, 21, 26, 28, 35, 36, 49, 72, 122, 142, 157, 161, 178, 183, 186–189, 202–204, 211–213, 215, 216, 218–220
Knowledge translation, 8, 18, 176, 177, 212

L

Lalonde report, 29
Lasswell, 25, 31
Latvia, 60
Life cycle, 7, 28, 33, 173, 177, 194
Lifelong learning, 183, 184, 197
Life trajectory, 174, 183, 184, 194
Lived [experiences], 17, 31, 187, 213
Living Experiences, 17
London, 129
Loneliness, 57, 62, 70, 71, 73, 77, 196
Long Term Care (LTC), 7, 57, 64, 65, 104
Long-term care systems, 15
Long term residential care, 8, 57, 59, 64, 143, 182
Low income countries, 53
LTC homes, 57, 59–61, 63, 64, 66–70, 73, 77, 96, 102, 144
LTC residents, 57, 59, 61–63, 66, 67, 69–73, 75–78, 163, 187

M

Marginalization, 151, 153, 154, 156, 188, 189, 214
Marginalized, 153, 214
Markov decision process, 75
McKinsey & Company, 99, 208
McMaster Health Forum, 65, 66, 187
MDS/RAI assessment systems, 68
Mechanical interventions, 50
Medical model, 29, 159
Medical model of intervention, 29, 153
Medicare, 34
MedTech Europe, 165
MEDTEQ+, 183
Mental health, 97, 102, 141, 153, 162
MHealth, 125, 128, 129, 215
Mieke van der Bijl-Brouwer, 32
Minimum Data Set (MDS), 68
Mobile health, 67, 76, 158, 215
Mobile phones, 13, 45, 50, 128, 158, 205
Mobility, 12, 72, 73, 77, 91–93, 99, 105, 119, 178
Models of care delivery, 125, 130
Monique Bégin, 34
Monitoring, 19, 47, 49, 51, 54, 65, 70, 76, 77, 121, 122, 125, 126, 142, 146, 157–159, 197, 198, 212, 220
Monitoring devices, 128
Morbidities, 15, 22, 31, 163, 181, 204, 215
Mortality, 18, 19, 31, 57, 61, 184
Multimorbidity, 15, 18, 142
Multiple Chronic Conditions (MCC), 20, 94

N

National Health Service (NHS), 127, 129
Neurological, 100, 206
New York, 72
New Zealand, 27, 101
North America, 125
Norway, 27, 34, 100

O

2021 OECD survey, 144
Office of the Seniors Advocate in British Columbia, 163
Ontario, 43, 51, 59, 100, 201, 210

Organisation for Economic Co-operation and Development (OECD), 60, 61, 63, 184

P

Pan-Canadian Health Data Strategy, 145
Pandemic, 3, 28, 42, 51, 57, 59, 61, 62, 64, 66, 67, 70, 71, 73, 77, 95, 105, 109, 126, 127, 145, 155, 156, 158, 159, 163, 165, 167, 176, 177, 184–186, 189, 198
Paradigm freeze, 34
Parkinson's disease, 50, 100
Paro, 74, 75
Path dependency, 5, 173, 177, 193, 215
Patient journey mapping, 21
Pension systems, 15
Personal Digital Assistants (PDAs), 128
Personal health record, 160, 161
Person-centered emotion-focused model of dementia care, 60
Pew Research Center report (2021), 19, 21, 27, 35, 36, 43, 50, 53, 60–67, 69–71, 73, 74, 76–78, 93–96, 98, 100, 101, 105, 121, 123, 126–129, 141–144, 146, 152, 158, 160, 163–165, 184, 187, 188, 202, 203, 205, 211, 215, 219, 220
Platform, 71, 73, 141, 142, 156, 159, 168, 177, 209
Poland, 60
Policy environment, 5
Policymaking, 7, 28, 29, 32, 159
Population, 5, 15–17, 21, 29–31, 49, 57, 59–62, 78, 94, 97, 98, 105, 108, 110, 121, 123, 124, 130, 153, 154, 157, 160–162, 166–168, 181, 183, 193, 201, 210, 215, 217
Population aging, 18
Population health approach, 29, 159
Portals, 103, 125, 128, 140, 142, 143, 145, 160, 221
Poverty, 29, 30, 110, 151–168, 182, 193, 195, 198
Primary care, 5, 7, 8, 11, 97, 119–131, 141–146, 151, 165, 167, 174, 181, 182, 185, 206, 216
Primary care setting, 22, 124, 126, 128, 141, 143
Primary disease prevention, 5
Privacy, 42, 43, 46, 54, 66, 74, 76, 129, 142, 143, 145, 160–162, 164, 166, 177, 186, 195
Professor Denise M. Rousseau, 36
Public health, 19, 28, 31, 44, 59, 108, 109, 127, 128, 145, 159, 174, 218

Q

Quality of care, 17, 60, 61, 69, 76, 78, 142, 145, 164, 166
Quality-of-life, 5–7, 11, 14, 17–19, 29, 31, 32, 37, 52–54, 57, 66, 71, 72, 76, 77, 91, 94, 97, 101, 104, 142, 153, 165, 175–178, 185, 193
Quick Response codes (QR), 50

R

Race, 62, 127, 211
Racism, 26, 62, 158, 211
RAI/MDS, 68
Rehabilitation, 5, 7, 8, 13, 41, 43, 47, 91–110, 119, 139, 140, 146, 174, 182, 210, 215, 216
Remote, 45, 49, 51, 54, 65, 74, 76, 77, 125–127, 158, 167, 177, 207, 220
Research, 5–8, 13, 17, 18, 21, 26, 27, 33, 37, 43, 44, 60, 62, 64, 70–74, 76, 98, 100, 102, 105, 108, 122, 126–129, 139, 142, 153, 155, 157, 158, 160, 162, 163, 172, 176, 177, 181, 183, 188, 189, 195, 196, 204, 208, 210, 212–215, 217–220
Researchers, 5, 8, 18, 20, 21, 29, 33–35, 37, 54, 68–70, 74, 78, 107, 155, 158, 159, 162–166, 168, 176–178, 183, 184, 186, 188, 212, 213
Research New Brunswick, 21
Research-Oriented Managed Access (ROMA), 33
Research Roundup, 21
Residential care, 59, 185
Residential care facilities, 142, 184
Reverse Trade Show, 106, 108
Robert Butler, 26

Robotics, 50, 74, 78, 94, 97–102, 108, 110
Robots, 65, 74–76, 98, 100, 101, 105, 110, 187

S
Safety, 6, 13, 14, 21, 27, 42, 47, 48, 62, 77, 120, 127, 141, 145, 151, 152, 158, 163, 164, 166, 175, 177, 178, 198, 209
SARS-Cov-2, 184
Security, 3, 22, 29, 42, 47–49, 51, 142, 145, 151, 160, 164, 177, 201
Self-care education, 19
Self-care support, 19, 185, 186
Self-management of chronic pain (SELMA), 128
Self-monitoring, 19
Senses, 11–15, 44, 178
Sensitive, 47, 76, 175, 176, 188
Sensitivity, 13, 158, 168
Sexism, 26
Sexual orientation, 62, 211, 214
Shiny Object Syndrome (SOS), 5, 8, 14, 30, 31, 71, 93, 106, 126, 144, 145, 160, 163, 165, 167, 168, 175, 187, 195, 205, 209, 210, 215, 216
Skeletomuscular functions, 15, 44
Skeletomuscular system, 15
Smart device, 158, 165, 168, 198
Smell, 14, 44
Social care, 20, 22, 29, 43, 57, 159, 165, 168, 174, 175, 177, 195, 196
Social determinants, 21, 154, 218
Social Determinants Of Health (SDOH), 29, 218, 219
Social isolation, 57, 62, 67, 70, 73, 77, 91, 98, 196
Social media, 71, 156, 164, 194, 205
Social model of disability, 29
Social needs, 7, 174, 219
Social robots, 65, 70, 73, 75–77, 187
Social security, 15, 27
South Korea, 69
Spain, 157
Spinal cord, 99, 100
Strategic Innovation Fund, 184
Strokes, 43, 92–96, 99–102, 104, 105, 201, 206
Structural determinants, 18, 20, 218
Subpopulations, 68, 152–154
Sustainability, 15, 34, 77, 129, 146, 173, 174
Sustainable Development Goal (SDG) 3.8, 145
Sweden, 27, 128
Switzerland, 27, 128
Sydney, 152
Systems for Person-centered Elder Care (SPEC), 69

T
Taste, 14, 44, 178
Technological barriers, 51, 66
Technology, 5, 6, 8, 12–14, 18, 19, 28, 30, 31, 33, 37, 43–48, 50, 51, 53, 54, 57, 64–67, 69–71, 73, 74, 76–78, 94, 95, 97–110, 120, 124–131, 141, 142, 145, 146, 155, 158, 163, 165, 172, 173, 175–178, 183, 185–189, 205, 208–210, 215, 221
Technology assessment issues, 8
Technology-supported integrated care model, 69
Teleconsultation, 125, 167, 185
Telehealth, 51, 125, 126, 157
Telemedicine, 51, 105, 125, 129, 177, 185
TELUS, 51, 129, 155
*The future of digital spaces and their role in demo*cracy, 164
The Netherlands, 27, 34, 59, 60, 145, 160
Dr. Theresa Tam, 19
Touch, 13, 30, 44, 46–48, 58, 209
Train-the-trainer, 69
Transdisciplinary research, 18
Turkey, 60
Type II diabetes, 11

U
UN Decade of Healthy Aging (2021–2030), 3
United Kingdom, 27, 51, 60, 94, 96, 121, 126, 127, 129
United Nations (UN), 3, 15, 16, 26, 185, 203
United Nations Children's Fund (UNICEF), 121
United Nation's Declaration of Human Rights affords, 157
United Nations Educational, Scientific, and Cultural Organization (UNESCO), 36
United Nations General Assembly, 53

United Nations Population Division, 15
United States, 27, 68, 94, 95, 99, 141
Universal health care, 15, 29, 51, 157
Usability, 47, 51, 67, 102, 130, 162, 186, 189, 195, 220

V

Value-of-information methods, 33
Virtual digital, 48, 49, 51
Virtual Reality (VR), 12, 67, 94, 97, 101–103, 107, 108, 188
Virtual Reverse Trade Show, 109
Vision, 7, 75, 92, 183
Visual acuity, 13, 47

W

Ward of the 21st Century (W21C) Research and Innovation Centre, 108
Welfare systems, 60
Wellness, 18, 95, 128, 159, 197
World Health Organization (WHO), 3, 29, 53, 94–96, 100, 108, 110, 121, 128, 130, 141, 156, 207, 211, 212, 214

MIX
Papier aus verantwortungsvollen Quellen
Paper from responsible sources
FSC® C105338

If you have any concerns about our products,
you can contact us on
ProductSafety@springernature.com

In case Publisher is established outside the EU,
the EU authorized representative is:
Springer Nature Customer Service Center GmbH
Europaplatz 3, 69115 Heidelberg, Germany

Printed by Libri Plureos GmbH
in Hamburg, Germany